Springer Finance

Springer Finance

Springer Finance is a programme of books aimed at students, academics and practitioners working on increasingly technical approaches to the analysis of financial markets. It aims to cover a variety of topics, not only mathematical finance but foreign exchanges, term structure, risk management, portfolio theory, equity derivatives, and financial economics.

Ammann M., Credit Risk Valuation: Methods, Models, and Application (2001)
Back K., A Course in Derivative Securities: Introduction to Theory and Computation (2005)
Barucci E., Financial Markets Theory. Equilibrium, Efficiency and Information (2003)
Bielecki T.R. and Rutkowski M., Credit Risk: Modeling, Valuation and Hedging (2002)
Bingham N.H. and Kiesel R., Risk-Neutral Valuation: Pricing and Hedging of Financial Derivatives (1998, 2nd ed. 2004)
Brigo D. and Mercurio F., Interest Rate Models: Theory and Practice (2001)
Buff R., Uncertain Volatility Models-Theory and Application (2002)
Carmona R.A. and Tehranchi M.R., Interest Rate Models: an Infinite Dimensional Stochastic Analysis Perspective (2006)
Dana R.A. and Jeanblanc M., Financial Markets in Continuous Time (2002)
Deboeck G. and Kohonen T. (Editors), Visual Explorations in Finance with Self-Organizing Maps (1998)
Delbaen F. and Schachermayer W., The Mathematics of Arbitrage (2005)
Elliott R.J. and Kopp P.E., Mathematics of Financial Markets (1999, 2nd ed. 2005)
Fengler M.R., Semiparametric Modeling of Implied Volatility (200)
Geman H., Madan D., Pliska S.R. and Vorst T. (Editors), Mathematical Finance–Bachelier Congress 2000 (2001)
Gundlach M., Lehrbass F. (Editors), CreditRisk$^+$ in the Banking Industry (2004)
Kellerhals B.P., Asset Pricing (2004)
Külpmann M., Irrational Exuberance Reconsidered (2004)
Kwok Y.-K., Mathematical Models of Financial Derivatives (1998)
Malliavin P. and Thalmaier A., Stochastic Calculus of Variations in Mathematical Finance (2005)
Meucci A., Risk and Asset Allocation (2005)
Pelsser A., Efficient Methods for Valuing Interest Rate Derivatives (2000)
Prigent J.-L., Weak Convergence of Financial Markets (2003)
Schmid B., Credit Risk Pricing Models (2004)
Shreve S.E., Stochastic Calculus for Finance I (2004)
Shreve S.E., Stochastic Calculus for Finance II (2004)
Yor M., Exponential Functionals of Brownian Motion and Related Processes (2001)
Zagst R., Interest-Rate Management (2002)
Zhu Y.-L., Wu X., Chern I.-L., Derivative Securities and Difference Methods (2004)
Ziegler A., Incomplete Information and Heterogeneous Beliefs in Continuous-time Finance (2003)
Ziegler A., A Game Theory Analysis of Options (2004)

René A. Carmona · Michael R. Tehranchi

Interest Rate Models: an Infinite Dimensional Stochastic Analysis Perspective

With 12 Figures

Springer

René A. Carmona
Bendheim Center for Finance
Department of Operations Research
and Financial Engineering
Princeton University
Princeton, NJ 08544
USA
E-mail: rcarmona@princeton.edu

Michael R. Tehranchi
Statistical Laboratory
Centre for Mathematical Sciences
University of Cambridge
Wilberforce Road
Cambridge CB3 0WB
UK
E-mail: m.tehranchi@statslab.cam.ac.uk

Mathematics Subject Classification (2000): 46T05, 46T12, 60H07, 60H15, 91B28
JEL Classification: E43, G12, G13

ISBN:13 978-3-642-06600-9 e-ISBN:13 978-3-540-27067-6

Springer is a part of Springer Science+Business Media
springer.com

Softcover reprint of the hardcover 1st edition 2006

Cover design: *design & production*, Heidelberg

Printed on acid-free paper 41/3100YL - 5 4 3 2 1 0

To Lenny Gross

Preface

The level of complexity of the bond market is higher than for the equity markets: one simple reason is contained in the fact that the underlying instruments on which the derivatives are written are more sophisticated than mere shares of stock. As a consequence, the mathematical models needed to describe their time evolution will have to be more involved. Indeed on each given day t, instead of being given by a single number S_t as the price of one share of a common stock, the term structure of interest rates is given by a curve determined by a finite discrete set of values. This curve is interpreted as the sampling of the graph of a function $T \hookrightarrow P(t, T)$ of the date of maturity of the instrument. In particular, whenever we have to deal with stock models involving ordinary or stochastic differential equations or finite dimensional dynamical systems, we will have to deal with stochastic partial differential equations or infinite dimensional systems!

The main goal of the book is to present, in a self-contained manner, the empirical facts needed to understand the sophisticated mathematical models developed by the financial mathematics community over the last decade. So after a very elementary introduction to the mechanics of the bond market, and a thorough statistical analysis of the data available to any curious spectator without any special inside track information, we gradually introduce the mathematical tools needed to analyze the stochastic models most widely used in the industry. Our point of view has been strongly influenced by recent works of Cont and his collaborators and the Ph.D. of Filipović. They merge the original proposal of Musiela inviting us to rewrite the HJM model as a stochastic partial differential equation, together with Björk's proposal to recast the HJM model in the framework of stochastic differential equations in a Banach space. The main thrust of the book is to present this approach from scratch, in a rigorous and self-contained manner.

Quick Summary. The first part comprises two chapters. The first one is very practical. Starting from scratch, it offers a lowbrow presentation of the bond markets. This chapter is of a descriptive nature, and it can be skipped by readers familiar with the mechanics of these markets or those who are more interested in mathematical models. The presentation is self-contained and no statistical prerequisites are needed despite the detailed discussion of Principal Component Analysis (PCA for short) and curve estimation. On the other hand, the discussion of the factor models found in the following chapter assumes some familiarity with Itô's stochastic calculus and the rudiments of the Black–Scholes pricing theory. This first part of the book constitutes a good introduction to the fixed income markets at the level of a Master in Quantitative Finance.

The second part is a course on infinite dimensional analysis. If it weren't for the fact that the choice of topics was motivated by the presentation of the stochastic partial differential and random field approaches to fixed income models, this part could be viewed as stand-alone text. Prompted by issues raised at the end of Part I, we introduce the theory of infinite dimensional Itô processes, and we develop the tools of infinite dimensional stochastic analysis (including Malliavin calculus) for the purpose of the general fixed income models we study in Part III of the book.

The last part of the book resumes the analysis of fixed income market models where it was left off at the end of Part I. The dynamics of the term structure are recast as a stochastic system in a function space, and the results of Part II are brought to bear to analyze these infinite dimensional dynamical systems both from the geometric and probabilistic point of views. Old models are revisited and new financial results are derived and explained in the light of infinitely many sources of randomness.

Acknowledgments. The first version of the manuscript was prepared as a set of lecture notes for a graduate seminar given by the first-named author during the summer of 2000 at Princeton University. Subsequently, the crash course on the mechanics of the bond market was prepared in December of 2000 for the tutorial presented in Los Angeles at the IPAM on the 3rd, 4th and 5th of January 2001. Rough drafts of the following chapters were added in preparation for short courses given in Paris in January 2001 and Warwick in March of the same year. RC would like to thank Jaksa Cvitanic, Nizar Touzi, and David Elworthy respectively, for invitations to offer these short courses. MT acknowledges support during writing of this book from the National Science Foundation in the form of a VIGRE postdoctoral fellowship. He would also like to thank Thaleia Zariphopoulou for inviting him to give a course on this material at the University of Texas at Austin during the fall of 2003.

Finally, we would like to dedicate this book to Leonard Gross. The footprints of his seminal work on infinite dimensional stochastic analysis can be found all over the text: the depth of his contribution to this corner of mathematics cannot be emphasized enough. And to make matter even more personal, RC would like to acknowledge an unrepayable personal debt to L. Gross for being an enlightening teacher, an enjoyable advisor, a role model for his humorous perspective on academia and life in general, and for being a trustworthy friend.

Princeton, NJ, October 2005 *René A. Carmona*
Cambridge, UK, October 2005 *Michael R. Tehranchi*

Contents

Part I The Term Structure of Interest Rates

Part III Generalized Models for the Term Structure

Part I

The Term Structure of Interest Rates

1

Data and Instruments of the Term Structure of Interest Rates

The size and the level of sophistication of the market of fixed income securities increased dramatically over the last twenty years and it became a prime test bed for financial institutions and academic research. The fundamental object to model is the term structure of interest rates, and we shall approach it via the prices of Treasury bond issues. Models for these prices are crucial for several reasons including the pricing of derivatives such as swaps, quantifying and managing financial risk, and setting monetary policy. We mostly restrict ourselves to Treasury issues to avoid credit issues and the likelihood of default.
We consider some of the fundamental statistical challenges of the bond markets after presenting a crash course on the mechanics of interest rates and the fixed income securities through which we gradually introduce concepts of increasing level of sophistication. This also gives us a chance to introduce the notation and the terminology used throughout the book.

1.1 Time Value of Money and Zero Coupon Bonds

We introduce the *time value of money* by valuing the simplest possible fixed income instrument. Like for all the other financial instruments considered in this book, we define it by specifying its cash flow. In the present situation, the instrument provides a single payment of a fixed amount (the principal or nominal value X) at a given date in the future. This date is called the maturity date. If the time to maturity is exactly n years, the present value of this instrument is:

$$P(X, n) = \frac{1}{(1+r)^n} X. \tag{1.1}$$

This formula gives the present value of a nominal amount X due in n years time. Such an instrument is called a *discount bond* or a *zero coupon bond* because the only cash exchange takes place at the end of the life of the instrument, i.e. at the date of maturity. The positive number r is referred to as the (yearly) *discount rate* or *spot interest rate* for time to maturity n since it is the interest rate which is applicable today (hence the terminology *spot*) on an n-year loan. Formula (1.1) is the simplest way to quantify the adage: *one dollar is worth more today than later!*

1.1.1 Treasury Bills

Zero coupon bonds subject to pricing formula (1.1) do exist. Examples include Treasury bills (T-bills for short) which are securities issued by the US government with a time to maturity of one year or less. A noticeable difference with the other securities discussed later is the fact that *they do not carry coupon payments.*

Let us consider for example the case of an investor who buys a $ 100000 13-week T-bill at a 6% yield (rate). The investor pays (approximately) $ 98500 at the inception of the contract, and receives the nominal value $ 100000 at maturity 13 weeks later. Since $13 = 52/4$ weeks represent one quarter, and since 6% is understood as an annual rate, the discount is computed as $100000 \times .06/4 = 1500$.

So in order to price a 5.1% rate T-bill which matures in 122 days, we first compute the discount rate:

$$\delta = 5.1 \times (122/360) = 1.728$$

which says that the investor receives a discount of $ 1.728 per $ 100 of nominal value. Consequently, the price of a T-bill with this (annual) rate and time to maturity should be:

$$\$\ 100 - \delta = 100 - 1.728 = 98.272$$

per $ 100 of nominal value.

Rates, yields, spreads, etc. are usually quoted in *basis points.* There are 100 basis points in one percentage point. The Treasury issues bills with times to maturity of 13 weeks, 26 weeks and 52 weeks. These bills are called *three-month bills, six-month bills* and *one-year bills,* although these names are accurate only at their inception. Thirteen-week bills and twenty-six week bills are auctioned off every Monday while the fifty-two week bills are auctioned off once a month.

The T-bill market is high volume, and liquidity is not an issue. We reproduce below the market quotes for the US government Treasury bills for Friday September 2nd, 2005 (source: eSpeed/Cantor Fitzgerald).

Treasury Bills

Maturity	Days to Mat.	Bid	Ask	Chg	Asked Yield
Sep 08 05	2	3.25	3.24	+0.06	3.29
Sep 15 05	9	3.46	3.45	-0.01	3.50
Sep 22 05	16	3.32	3.31	-0.01	3.36
Sep 29 05	23	3.32	3.31	+0.02	3.36

Oct 06 05	30	3.27	3.26	+0.01	3.31
Oct 13 05	37	3.27	3.26	+0.02	3.32
Oct 20 05	44	3.27	3.26	+0.01	3.32
Oct 27 05	51	3.26	3.25	-0.01	3.31
Nov 03 05	58	3.31	3.30	+0.02	3.36
Nov 10 05	65	3.32	3.31	....	3.38
Nov 17 05	72	3.34	3.33	+0.01	3.40
Nov 25 05	80	3.36	3.35	+0.01	3.42
Dec 01 05	86	3.38	3.37	+0.01	3.44
Dec 08 05	93	3.41	3.40	+0.02	3.48
Dec 15 05	100	3.41	3.40	+0.02	3.48
Dec 22 05	107	3.39	3.38	-0.02	3.46
Dec 29 05	114	3.42	3.41	-0.01	3.50
Jan 05 06	121	3.44	3.43	+0.02	3.52
Jan 12 06	128	3.45	3.44	+0.01	3.53
Jan 19 06	135	3.46	3.45	+0.01	3.54
Jan 26 06	142	3.47	3.46	+0.01	3.56
Feb 02 06	149	3.48	3.47	....	3.57
Feb 09 06	156	3.48	3.47	-0.01	3.57
Feb 16 06	163	3.50	3.49	+0.01	3.60
Feb 23 06	170	3.49	3.48	-0.01	3.59
Mar 02 06	177	3.51	3.50	....	3.61

The first column gives the date of maturity of the bill, while the second column gives the number of days to maturity. See our discussion of day count conventions later in this chapter. The third and fourth columns give the bid and ask prices in decimal form. Compare with the bid and ask columns of the quotes we give below for Treasury notes and bonds on the same day, and the ensuing discussion.

1.1.2 Discount Factors and Interest Rates

Since the nominal value X appears merely as a plain multiplicative factor in formula (1.1), it is convenient to assume that its value is equal to 1, and effectively drop it from the notation. This leads to the notion of discount factor. Discount factors can be viewed as quantities used at a given point in time to obtain the present value of future cash flows. At a given time t, the discount factor $P_{t,m}$ with time to maturity m, or maturity date $T = t + m$, is given by the formula:

$$P_{t,m} = \frac{1}{(1 + r_{t,m})^m} \tag{1.2}$$

where $r_{t,m}$ is the yield or yearly spot interest rate in force at time t for this time to maturity. We assumed implicitly that the time to maturity $T - t$ is a whole number m of years. Definition (1.2) can be rewritten in the form:

$$\log(1 + r_{t,m}) = -\frac{1}{m} \log P_{t,m}$$

and considering the fact that $\log(1 + x) \sim x$ when x is small, the same definition gives the approximate identity:

$$r_{t,m} \sim -\frac{1}{m} \log P_{t,m}$$

which becomes an exact equality if we use continuous compounding. This formula justifies the terminology discount rate for r. Considering payments occurring in m years time, the spot rate $r_{t,m}$ is the single rate of return used to discount all the cash flows for the discrete periods from time t to time $t+m$. As such, it appears as some sort of composite of interest rates applicable over shorter periods. Moreover, this formula offers a natural generalization to continuous time models with continuous compounding of the interest. As we shall see later, this extension reads:

$$P(t,T) = e^{-(T-t)r(t,T)}. \tag{1.3}$$

where $P(t,t+m) = P_{t,m}$ and $r(t,t+m) = r_{t,m}$.

The discount factor is a very useful quantity. Indeed, according to the above discussion, the present value of any future cash flow can be computed by multiplying its nominal value by the appropriate value of the discount factor. We use the notation $P(t,T)$ to indicate that is the price at time t of a zero coupon bond with maturity T.

The information contained in the graph of the discount factor as a function of the maturity T (i.e the so-called discount function) is often repackaged in quantities which better quantify the returns associated with purchasing future cash flows at their present value. These quantities go under the names of *spot-interest-rate* curve, *par-yield* curve, and *implied forward rate* curve. This chapter is devoted to the introduction of these quantities in the discrete time setting, and to the definition of their analogs in the continuous time limit. The latter is a mathematical convenience which makes it possible to use the rules of the differential and integral calculus. It is somehow unrealistic because money is lent for discrete periods of time, but when these periods are short, the continuous time limit models become reasonable. We shall discuss later in the next chapter how to go from discrete data to continuous time models and vice versa.

1.2 Coupon Bearing Bonds

Now that we know what a zero coupon bond is, it is time to introduce the notion of coupon bearing bond. If a zero coupon bond was involving only one payment, what is called a bond (or a coupon bearing bond), is a regular stream of future cash flows of the same type. To be more specific, a *coupon bond* is a series of payments amounting to C_1, C_2, ... , C_m, at times T_1, T_2, ... , T_m, and a terminal payment X at the maturity date T_m. The coupon payments are in arrears in the sense that at date T_j, the coupon payment is the reward for the interests accrued up until T_j. As before, X is called the nominal value, or the face value, or the principal value of the bond. According to the above discussion of the discount factors, the bond price at time t should be given by the formula:

$$B(t) = \sum_{t \le T_j} C_j P(t,T_j) + XP(t,T_m). \tag{1.4}$$

This all-purpose formula can be specialized advantageously for in most cases, the payments C_j are made at regular time intervals. Also, these coupon payments C_j are most often quoted as a percentage c of the face value X of the bond. This percentage is given as an annual rate, even though payments are sometimes made every six months, or at some other interval. It is convenient to introduce a special notation, say n_y, for the number of coupon payments in one year. For example, $n_y = 2$ for coupons paid semi-annually. In this notation, the coupon payment is expressed as $C_j = cX/n_y$. If we denote by $r_1, r_2, \ldots, r_m$ the interest rates for the m periods ending with the coupon payments $T_1, T_2, \ldots, T_m$, then the present value of the bond cash flow is given by the formula:

$$\begin{aligned} B(t) &= \frac{C_1}{1+r_1/n_y} + \frac{C_2}{(1+r_2/n_y)^2} + \cdots + \frac{C_m}{(1+r_m/n_y)^m} + \frac{X}{(1+r_m/n_y)^m} \\ &= \frac{cX}{n_y(1+r_1/n_y)} + \frac{cX}{n_y(1+r_2/n_y)^2} + \cdots + \frac{(1+c/n_y)X}{(1+r_m/n_y)^m}\,. \end{aligned} \tag{1.5}$$

Note that we divided the rates r_n by the frequency n_y because the rates are usually quoted as yearly rates. Formulae (1.4) and (1.5) are often referred to as the *bond price equations.* An important consequence of these formulae is the fact that on any given day, the value of a bond is entirely determined by the sequence of yields r_n, or equivalently by the discount factors $P(t, t+n)$ which form a sampling of the discount curve.

Remarks.

1. Reference to the present date t will often be dropped from the notation when no confusion is possible. Moreover, instead of working with the absolute dates $T_1, T_2, \ldots, T_m$, which can represent coupon payment dates as well as maturity dates of various bonds, it will be often more convenient to work with the times to maturities which we denote by $x_1 = T_1 - t$, $x_2 = T_2 - t$, $\ldots$, $x_m = T_m - t$. We will use whatever notation is more convenient for the discussion at hand.
2. Unfortunately, bond prices are not quoted as a single number. Instead, they are given by a *bid–ask* interval. We shall ignore the existence of this bid–ask spread for most of what follows, collapsing this interval to a single value, by considering its midpoint for example. We shall reinstate the bid–ask spread when we discuss the actual statistical estimation procedures.

1.2.1 Treasury Notes and Treasury Bonds

Treasury notes are Treasury securities with time to maturity ranging from 1 to 10 years at the time of sale. Unlike bills, they have coupons: they pay interest every six months. Notes are auctioned on a regular cycle. The Fed acts as the agent for the Treasury, awarding competitive bids in decreasing order of price, highest prices first. The smallest nominal amount is \$ 5000 for notes with two to three years to maturity at the time of issue, and it is $ 1000 for notes with four or more years to maturity at the time of issue. Both types are available in multiples of $ 1000 above the minimum nominal amount.

Treasury bonds, or T-bonds, are Treasury securities with more than 10 years to maturity at the time of sale. Like Treasury notes, they are sold at auctions, they are traded on a dollar price basis, they bear coupons and they accrue interest. Apart from their different life spans, the differences between Treasury notes and bonds are few. For example, bonds have a minimal amount of $ 1000 with multiples of $ 1000 over that amount. Also, some bonds can have a call feature (see Sect. 1.5 below). For the purpose of this book, these differences are irrelevant: we shall give the same treatment to notes and bonds, and we shall only talk about T-bills and T-bonds, implicitly assuming that prices have already being adjusted for these extra features. The following quote from *The Wall Street Journal* of July 29, 2005 gives an indication of the unofficial early notices given to potential investors by the press:

> *The Treasury plans to raise about $ 1.98 billion in fresh cash Monday with the sale of about $ 34 billion in short-term bills to redeem $ 32.02 billion in maturing bills. The amount is down from $ 36 billion sold the previous week.*
>
> *The offering will include $ 18 billion of 13-week bills, and $ 16 billion of 26-week bills, which will mature Nov. 3, 2005, and Feb. 2, 2006, respectively. The CUSIP number for the three-month bills is 912795VY4, and for the six-month bills is 912795WM9. Noncompetitive tenders for the bills, available in minimum denominations of $ 1,000, must be received by noon EDT on Monday, and competitive tenders by 1 p.m.*

On any given day, there is a great variety of notes and bonds outstanding, with maturity ranging from a few days to 30 years, and coupon rates as low as 1.5 and as high as 7.5%. We reproduce the beginning and the end of the quotation list of Over-the-Counter (OTC) quotation based on transactions of $ 1 million or more. Treasury bond and note quotes are from mid-afternoon, September 2nd, 2005 (source: eSpeed/Cantor Fitzgerald).

US Government Bonds and Notes

Rate	Maturity Mo Yr	Bid	Ask	Chg	Asked Yield
1 5/8	Sep 05 n	99:28	99:29	+1	3.04
1 5/8	Oct 05 n	99:23	99:24	+1	3.29
5 3/4	Nov 05 n	100:13	100:14	....	3.38
5 7/8	Nov 05 n	100:13	100:14	-1	3.42
1 7/8	Nov 05 n	99:19	99:20	....	3.42
1 7/8	Dec 05 n	99:14	99:15	+1	3.57
1 7/8	Jan 06 n	99:09	99:10	....	3.57
...	...	...	...	...	...
...	...	...	...	...	...
5 1/4	Nov 28	113:03	113:04	-2	4.34
5 1/4	Feb 29	113:07	113:08	-3	4.34
3 7/8	Apr 29 i	139:13	139:14	-10	1.81
6 1/8	Aug 29	126:14	126:15	-2	4.34
6 1/4	May 30	128:29	128:30	-3	4.33
5 3/8	Feb 31	116:18	116:19	-4	4.30
3 3/8	Apr 32 i	135:00	135:01	-11	1.73

The first column gives the rate while the second column gives the month and year of maturity, with a lower case letter "n" where the instrument is a note and a lower case "i" if the issue is inflation indexed. See the discussion of indexed securities later in the chapter. The third and fourth columns give the bid and asked prices. Notice that none of the decimal parts happens to be greater than 31. Compare with the T-bill quotes given earlier. The reason is that the prices of Treasury notes and bonds are quoted in percentage points and 32nds of a percentage point. These are percentages of the nominal amount. But even though the figures contain a decimal point, the numbers to the right of the decimal point give the number of 32nds. So the first bid price, which reads 99.28, is actually $99 + 28/32 = 99.875$, which represents \$ 998750 per million of dollars of nominal amount. The fifth column gives the change in asked price with the asked price of the last trading day while the last column gives the yield computed on the asked price. More needs to be said on the way this yield is computed for some of the bonds (for callable bonds for example) but this level of detail is beyond the scope of this introductory presentation.

Because of their large volume, Treasury notes and bonds can easily be bought and sold at low transaction cost. They pay interest semi-annually, most often on the anniversary of the date of issue. This income is exempt from state income taxes.

After retiring the 30-year bond in October 2001, the Treasury resumed the sale of its 30-year bond, the "long bond," as it is often referred to by traders, starting in February 2006.

1.2.2 The STRIPS Program

Formula (1.4) shows that a coupon bearing bond can be viewed as a composite instrument comprising a zero coupon bond with the same maturity T_m and face value $(1+c/n_y)X$, and a set of zero coupon bonds whose maturity dates are the coupon payment dates T_j for $1 \leq j < m$ and face value cX/n_y. This remark is much more than a mere mathematical curiosity. Indeed, the principal and the interest components of US Treasury bonds have been traded as separate zero coupon securities under the Treasury STRIPS (Separate Trading of Registered Interest and Principal Securities) program since 1985. The program was created to meet the demand for zero coupon obligations. They are not special issues: the Treasury merely declares that specific notes and bonds (and no others) are eligible for the STRIPS program, and the *stripping* of these issues is done by government securities brokers and dealers who give a special security identification number (CUSIP in the jargon of financial data) to these issues. Figure 1.1 shows how STRIPS are quoted daily (at least following each day the bond market is open for trading) in *The Wall Street Journal.*

STRIPS are popular for many reasons. They have predictable cash flows: they pay their face value at maturity as they are not callable. As backed

U.S. Treasury Strips

MATURITY		TYPE	BID	ASKED	CHG	ASK YLD
Oct	05	ci	99:31	00	...	0.00
Nov	05	ci	99:22	99:22	...	3.34
Nov	05	np	99:22	99:23	...	3.31
Nov	05	np	99:22	99:22	...	3.32
Jan	06	ci	99:04	99:05	...	3.33
Feb	06	ci	98:24	98:25	...	3.62
Feb	06	bp	98:26	98:26	...	3.49
Feb	06	np	98:22	98:23	1	3.81
Apr	06	ci	98:02	98:02	...	3.90
May	06	ci	97:23	97:23	...	3.94
May	06	np	97:20	97:21	...	4.07
May	06	np	97:20	97:21	...	4.06
Jul	06	ci	97:07	97:08	...	3.72
Aug	06	ci	96:22	96:23	...	4.02
Aug	06	np	96:18	96:19	...	4.18
Oct	06	ci	96:00	96:01	...	4.08
Nov	06	ci	95:20	95:21	...	4.12
Nov	06	np	95:17	95:18	...	4.21
Nov	06	np	95:17	95:18	...	4.21
Feb	07	ci	94:21	94:22	−1	4.12
Feb	07	np	94:21	94:21	−1	4.13
May	07	ci	93:20	93:21	−1	4.17
May	07	np	93:21	93:21	−1	4.16
May	07	np	93:19	93:19	−1	4.20
Aug	07	ci	92:20	92:21	−1	4.19
Aug	07	np	92:19	92:20	−1	4.21
Aug	07	np	92:19	92:20	−1	4.21

Fig. 1.1. *Wall Street Journal* STRIPS quotes on October 15, 2005.

by US Treasury securities, they are very high quality debt instruments: they carry essentially no default risk. Also, they require very little up-front capital investment. Indeed, while Treasury bonds require a minimum investment of \$ 10000, some STRIPS may only require a few hundred dollars. Finally, as they mostly come from interest payments, they offer an extensive range of maturity dates.

1.2.3 Clean Prices

Formulae (1.4) and (1.5) implicitly assume that t is the time of a coupon payment, and consequently, that the time to maturity is an integer multiple of the time separating two successive coupon payments. Because of the very nature of the coupon payments occurring at specific dates, the bond prices given by the bond pricing formula (1.4) are discontinuous: they jump at the times the coupons are paid. This is regarded as an undesirable feature, and systematic price corrections are routinely implemented to remedy the jumps. The technicalities behind these price corrections increase the level of complexity of the formulae, but since most bond quotes (whether they are from the US Treasury, or from international or corporate markets) are given in terms of these corrected prices, we thought that it would be worth our time looking into a standard way to adjust the bond prices for these jumps.

The most natural way to smooth the discontinuities is to adjust the bond price for the *accrued interest* earned by the bond holder since the time of the last coupon payment. This notion of accrued interest is quantified in the following way. Since the bond price jumps by the amount cX/n_y at the times T_j of the coupon payments, if the last coupon payment (before the present time t) was made on date T_n, then the interests accrued since the last payment should be given by the quantity:

$$AI(T_n, t) = \frac{t - T_n}{T_{n+1} - T_n} \frac{cX}{n_y}, \tag{1.6}$$

and the *clean price* of the bond is defined by the requirement that the transaction price be equal to the clean price plus the accrued interest. In other words, if $T_n \leq t < T_{n+1}$, the clean price $CP(t, T_m)$ is defined as:

$$CP(t, T_m) = B(t) - AI(t, T_n) \tag{1.7}$$

where $B(t)$ is the transaction price given by (1.4) with the summation starting with $j = n + 1$.

1.3 Term Structure as Given by Curves

1.3.1 The Spot (Zero Coupon) Yield Curve

The spot rate $r_{t,m}$ at time t for time to maturity m (also denoted $r(t, T)$ if the time of maturity is $T = t + m$) defined from the discount factor via formulae (1.2) or (1.3) is called the *zero coupon yield* because it represents the yield to maturity on a zero coupon bond (also called a *discount bond*). Given observed values P_j of the discount factor, these zero coupon yields can be computed by inverting formula (1.2). Dropping the date t from the notation, we get:

$$P_m = \frac{1}{(1 + r_m)^m} \quad \Longleftrightarrow \quad r_m = \left(\frac{1}{P_m^{1/m}} - 1 \right) \tag{1.8}$$

for the zero coupon yield. The sequence of spot rates $\{r_j\}_{j=1,\dots,m}$ where m is time to maturity is what is called the *term structure of (spot) interest rate* or the *zero coupon yield curve.* It is usually plotted against the time to maturity $T_j - t$ in years. Figure 1.2 shows how the Treasury yield curves are pictured daily in *The Wall Street Journal.* Notice the non-uniform time scale on the horizontal axis.

Instead of plotting the yield on a given day t as a function of the time to maturity $T_j - t$, we may want to consider the historical changes in t of the yield for a fixed time to maturity $T_j - t$. This is done in Fig. 1.3 for a 10-year time to maturity, using again *The Wall Street Journal* as source. This plots shows a totally different behavior. The graph looks more like a sample path of a (stochastic) diffusion process than a piecewise linear interpolation of a smooth curve. This simple remark is one of the stylized facts which we shall try to capture when we tackle the difficult problem of modeling the stochastic dynamics of the term structure of interest rates.

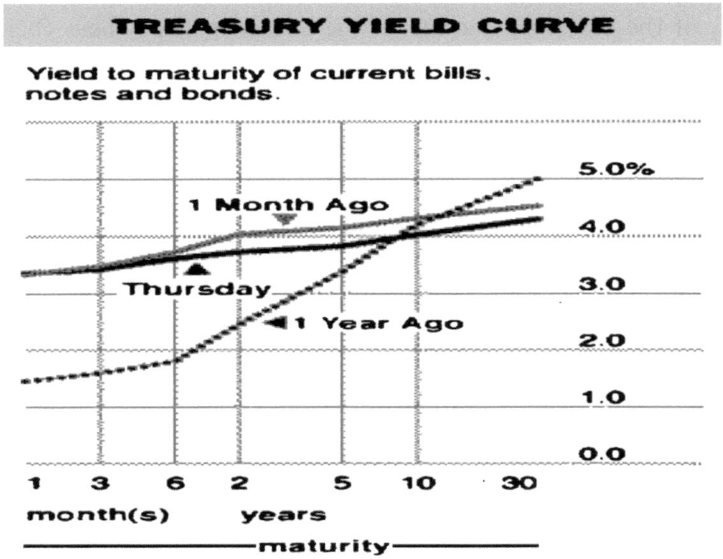

Fig. 1.2. Treasury yield curves published in *The Wall Street Journal* on September 1, 2005.

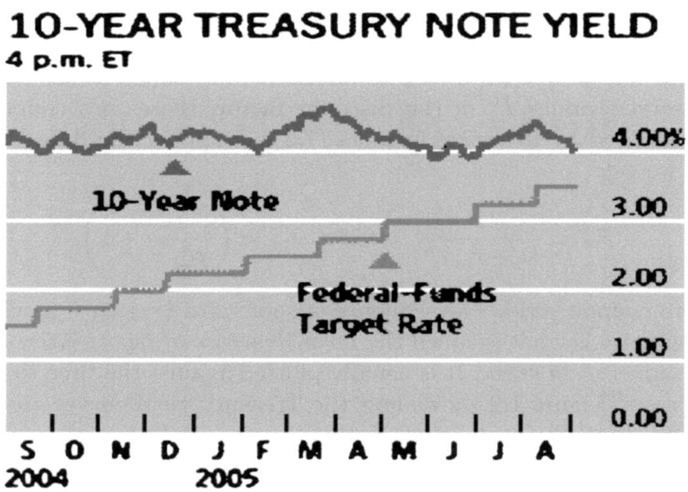

Fig. 1.3. Historical values of the Treasury 10 year yield as published in *The Wall Street Journal* on September 1, 2005.

The Par Yield Curve

The *par yield curve* has been introduced to give an account of the term structure of interest rates when information about coupon paying bonds is the only data available. Indeed, yields computed from coupon paying bonds can be quite different from the zero coupon yields computed as above.

A coupon paying bond is said to be priced *at par* if its current market price equals its face (or par) value. The yield of a coupon paid at par is equal to its coupon rate. This intuitive fact can be derived by simple manipulations after setting $y = c$ in the bond price equation. If the market price of a bond is less than its face value, we say the bond trades *at a discount*. In this case, its yield is higher than the coupon rate. If its market price is higher than its face value, it is said to trade *at a premium*. In this case the yield is lower than the coupon rate. These price/yield qualitative features are quite general: everything else being fixed, higher yields correspond to lower prices, and vice versa. Also note that, if the prices of two bonds are equal, then the one with the larger coupon has the highest yield.

The *par yield* is defined as the yield (or coupon) of a bond priced at par. In other words, if the bond is m time periods away from maturity, using the notation of formula (1.5), the *par yield* is the value of c for which we have the following equality:

$$1 = \sum_{j=1}^{m} \frac{c}{n_y(1 + r_j/n_y)^j} + \frac{1}{(1 + r_m/n_y)^m}.$$

1.3.2 The Forward Rate Curve and Duration

We restate the properties of the spot rate $r_{t,m}$ in terms of a new function for which we introduce the notation $f_{t,m}$. It is intended to represent the rate applicable from the end of the $(m-1)$-th period to the end of the m-th period. With this notation at hand we have:

$$\begin{aligned}
1/P_{t,1} &= 1 + r_{t,1} = 1 + f_{t,1} \\
1/P_{t,2} &= (1 + r_{t,2})^2 = (1 + f_{t,1})(1 + f_{t,2}) \\
\cdots &= \cdots \\
1/P_{t,j-1} &= (1 + r_{t,j-1})^{j-1} = (1 + f_{t,1})(1 + f_{t,2}) \cdots (1 + f_{t,j-1}) \\
1/P_{t,j} &= (1 + r_{t,j})^j = (1 + f_{t,1})(1 + f_{t,2}) \cdots (1 + f_{t,j-1})(1 + f_{t,j}).
\end{aligned}$$

Computing the ratio of the last two equations gives:

$$\frac{P_{t,j-1}}{P_{t,j}} = 1 + f_{t,j}$$

or equivalently:

$$f_{t,j} = \frac{P_{t,j-1} - P_{t,j}}{P_{t,j}} = -\frac{\Delta P_{t,j}}{P_{t,j}} \tag{1.9}$$

if we use the standard notation $\Delta P_{t,j} = P_{t,j} - P_{t,j-1}$ for the first difference of a sequence (i.e. the discrete time analog of the first derivative of a function). The rates $f_{t,1}, f_{t,2}, \dots, f_{t,j}$ implied by the discount factors $P_{t,1}, P_{t,2}, \dots, P_{t,j}$ are called the *implied forward interest rates.* The essential difference between the spot rate $r_{t,j}$ and the forward rate $f_{t,j}$ can be best restated by saying that $r_{t,j}$ gives the average rate of return of the next j periods while the forward rate $f_{t,j}$ gives the marginal rate of return over the j-th period, for example the one-year rate of return in 10-years time instead of today's 10 year rate.

We now define the notion of duration for a bond. If for a coupon bond with yield y, we denote by $C_1, C_2, \dots, C_m$ the coupons which take place at times $T_1, T_2, \dots, T_m$, the price B of the bond can be written as

$$B = \sum_{i=1}^{m} \frac{C_i}{(1 + y/n_y)^i}. \tag{1.10}$$

For the sake of notation, the present time (think of $t = 0$) is dropped from the notation, and we assume that the last coupon C_m includes the nominal payment X. With this notation, the Macaulay duration is defined as

$$D_M = \frac{1}{B} \sum_{i=1}^{m} T_i \frac{C_i}{(1 + y/n_y)^i}. \tag{1.11}$$

According to its definition, the duration is some form of expected time to payments. Indeed, it is a weighted average (with probability weights summing up to 1) of the coupon dates $T_1, \dots, T_m$, and for this reason it provides a *mean time to coupon payment.* It plays an important role in interest rate risk management. It is easy to see that the first derivative of the bond price B given by (1.10) with respect to the yield to maturity y is exactly the Macaulay duration D_M. So the duration measures (at least to first order) the sensitivity of the bond price to changes in the yield to maturity. As such, it is a crucial tool in the immunization of bond portfolios.

1.3.3 Swap Rate Curves

Swap contracts have been traded publicly since 1981. They are currently the most popular fixed income derivatives. Because of this popularity, the swap markets are extremely liquid, and as a consequence, they can be used to hedge interest rate risk of fixed income portfolios at a low cost.

Swap Contracts

As implied by its name, a swap contract obligates two parties to exchange (or swap) some specified cash flows at agreed-upon times. The most common swap contracts are interest rate swaps. In such a contract, one party, say counterparty A, agrees to make interest payments determined by an instrument P_A (say, a 30-year US Treasury bond rate), while the other party, say counterparty B, agrees to make interest payments determined by another instrument P_B (say, the London Interbank Offer Rate – LIBOR for short). Even though there are many variants of swap contracts, in a typical contract, the principal on which counterparty A makes interest payments is equal to the principal on which counterparty B makes interest payments. Also, the payment schedules are identical and periodic, the payment frequency being quarterly, semi-annually, etc.

It should be clear from the above discussion that a swap contract is equivalent to a portfolio of forward contracts, but we shall not use this feature here. In this section, we shall restrict ourselves to the so-called plain vanilla contracts involving a fixed interest rate and the 3- or 6-month LIBOR rate. See next section for the definition of these rates, and Chap. 7 for a first analysis of some of its derivatives.

A Price Formula for a Plain Vanilla Swap

Let us denote by X the common principal, by R the fixed interest rate on which the swap is written, by T_1, T_2, ... , T_m the dates after the current date t, at which the interest rate payments are scheduled, and by $L(t, T_{j-1})$ the LIBOR rate over the period $[T_{j-1}, T_j)$. On each payment date T_j, the variable interest rate used to compute the payment at this time is taken from the period $[T_{j-1}, T_j)$, so that the floating-interest payment for this period will be $X(T_j - T_{j-1})L(t, T_{j-1})$, while the fixed-interest payment for the same period will by $X(T_j - T_{j-1})R$. Such a contract is called a forward swap settled in arrears. Using the discount factors to compute the present value of the cash flows, we get:

$$P_{swap} = X \sum_{j=1}^{m} (T_j - T_{j-1})(L(t, T_{j-1}) - R)P(t, T_j)$$

where we use the convention $T_0 = t$. Notice that, if we were to add a payment of the principal X at time T_m, then the cash flows of the swap would be identical to the cash flows generated by a portfolio long a (fixed rate) coupon bearing bond and short a floating rate bond with the same face value. In the financial jargon, being long an instrument means having bought the instrument, while being short means having borrowed the instrument with the commitment to return it (with interest which we shall ignore here) and sold it.

The valuation problem for a swap is solved by computing the difference of the floating-rate bond and the fixed-coupon bond. Expressing the LIBOR in terms of discount factors (see next section), after simple algebraic manipulations we get:

$$P_{swap}(t, T) = X \left(1 - [P(t, T_m) + R \sum_{j=1}^{m} (T_j - T_{j-1})P(t, T_j)] \right) \tag{1.12}$$

where we use the standard notation $P(t,T)$ for the price at time t of a riskless zero coupon bond with maturity date T and nominal value 1.

The Swap Rate Curve

On any given day t, the swap rate $R_{swap}(t,T)$ with maturity $T = T_m$ is the unique value of the fixed rate R, which, once injected in formula (1.12) makes the swap price equal to 0. In other words, the swap rate is the value of the fixed interest rate for which the counterparties will agree to enter the swap contract without paying or receiving a premium. This swap rate is obtained by solving for R in the equation obtained by setting $P_{swap}(t,T) = 0$ in formula (1.12). This gives:

$$R_{swap}(t,T_m) = \frac{1 - P(t,T_m)}{\sum_{j=1}^{m}(T_j - T_{j-1})P(t,T_j)}. \tag{1.13}$$

Notice that in practice, the interest payments are regularly distributed over time (in other words the lengths of all the time intervals $T_j - T_{j-1}$ are equal), and for this reason, one often uses a parameter giving the frequency of the payments. Figure 1.4 shows that the curve of the swap rates is given in *The Wall Street Journal* with the same exposure as the yield curve. We shall come back to swap rates in Sect. 1.7 when we discuss principal component analysis.

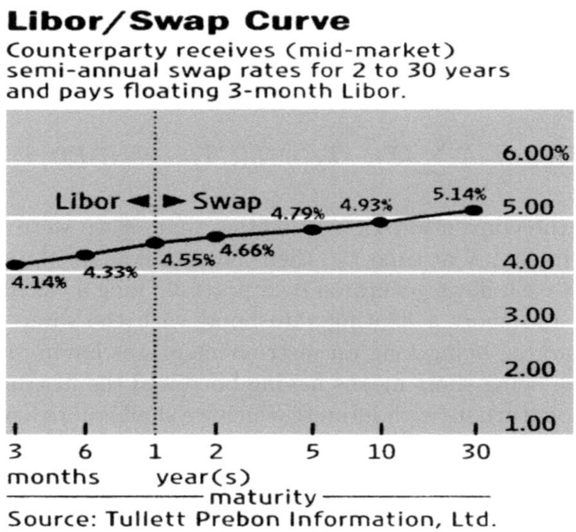

Fig. 1.4. Swap curve as published in *The Wall Street Journal* on 10/15/2005.

1.4 Continuous Compounding and Market Conventions

We now introduce the framework of continuous time finance which will be used throughout the book. The overarching assumption of the mathematical theory presented in this book is that at each time t, we have a continuum of liquidly traded zero coupon bonds, one for each possible date of maturity T after t. We shall eventually assume that dates are non-negative real numbers, and we shall assume the existence of a money-market account whose value B_t at time t is the result of investing a unit amount at time $t = 0$ in an account where money grows at the instantaneous rate r_t. In other words, B_t is the solution of the ordinary differential equation $dB_t = r_t B_t \, dt$ with initial condition $B_0 = 1$. The solution is given by

$$B_t = e^{\int_0^t r_s ds}.$$

The discount factor $D(t,T)$ between two dates t and T is defined at the amount at time t equivalent to unity at time T. Hence

$$D(t,T) = \frac{B_t}{B_T} = e^{-\int_t^T r_s ds}.$$

In this set-up a zero coupon bond with maturity T is a contract that guarantees its holder a unit payment at time T without any intermediate payment. We denote by $P(t,T)$ its value at time t. Notice that $D(t,T) = P(t,T)$ when $\{r_s\}_s$ is deterministic. However, they are different when $\{r_s\}_s$ is random. Indeed, $D(t,T)$ *randomness* depends upon the evolution of r_s for $t \leq s \leq T$ while $P(t,T)$ is known at time t. A form of the expectation hypothesis states that $P(t,T)$ is the expectation of $D(t,T)$.

1.4.1 Day Count Conventions

The reader must already have felt some discomfort with our loose treatment of time spans and dates, and it is about time to come clean and address some of the issues which we conveniently *brushed under the rug*. We consistently tried to use the letter t to denote the *present* date and the upper case T for the date of maturity of whatever instrument we are considering. In practice, these dates or times are given in a specific format, say:

$$t \leftrightarrow D_1 = (d_1/m_1/y_1)$$
$$T \leftrightarrow D_2 = (d_2/m_2/y_2)$$

where the dates D_i are given by their respective components, d_i for the day of the month, m_i for the month of the year, and y_i for the year. Day traders use high-frequency data, and in this case the date/time t is usually in the format

$$t \leftrightarrow D = (d/m/y\ H : M : S)$$

where H gives the hour of the day, m the number of minutes since the beginning of the hour and S the number of seconds. For the sake of simplicity we shall only discuss the case of times and dates given in the day/month/year format. Once a specific format is chosen for the dates and time, the main challenge is the computation of time intervals. Indeed, the time separating two given dates is what really matters the most for us, and the question to address is:

> *How should we compute a time to maturity $x = x(t, T)$ when t and T are dates given as above?*

Despite the obvious urge to use the notation $x = T - t$ for the length $x(t, T)$ of the time between t and T, we resist this temptation until the end of this section. The problem is solved by choosing a day count convention. We mention three of the most commonly used conventions. In all cases, the time span $x = T - t$ is measured in years.

- Actual/365: under this convention, $x = x(t, T)$ is given the value of the number of days between t and T divided by 365.
- Actual/360: under this convention, x is given the value of the number of days between t and T divided by 360.
- 30/360: under this convention, $x = x(t, T)$ is given the value of the number of days between t and T divided by 360 if we compute the number of days assuming that each month has exactly 30 days. To be specific, assuming that $T = (d_2/m_2/y_2)$ is a date coming after $t = (d_1/m_1/y_1)$, we set:

$$x = x(t, T) = y_2 - y_1 + \frac{(m_2 - m_1 - 1)^+}{12} + \frac{d_2 \wedge 30 + (30 - d_1)^+}{360}.$$

The prescriptions given in the bullet points above resolve most of the issues, except that there are still nagging problems to deal with, such as weekends, holidays, etc. which we shall not address here.

Financial markets quote interest rates, not zero coupon bonds. However, all quoted interest rates can be computed from zero coupon bond prices. Zero coupon bonds are more of a convenient mathematical construct rather than a quantity quoted in interbank transactions. In order to go from zero coupon prices to interest rates and back, we need two things:

- choosing a day count convention;
- choosing a compounding convention.

We already addressed the first bullet point. Building on our introductory discussion of discrete time examples, we now review the various continuous compounding conventions which we follow in this book.

1.4.2 Compounding Conventions

The continuously compounded spot interest rate prevailing at time t for maturity T is the constant rate $r = r(t,T)$ for which an investment of $P(t,T)$ at time t will produce a cash flow of 1 at maturity. In other words, $r(t,T)$ is the number r such that $e^{rx(t,T)}P(t,T) = 1$ or equivalently:

$$r(t,T) = -\frac{1}{x(t,T)} \log P(t,T). \tag{1.14}$$

We recast the *discrete time* interest rate manipulations done so far in the present mathematical framework of continuous time finance in the following way. The simply compounded spot interest rate prevailing at time t for maturity T is the constant rate $r = L(t,T)$ for which an investment of $P(t,T)$ at time t will produce a cash flow of 1 at maturity as above, when interests accrue proportionally to the length of time of the investment. In other words, $L(t,T)$ is the number r such that $(1 + rx(t,T))P(t,T) = 1$ or equivalently:

$$L(t,T) = \frac{1 - P(t,T)}{x(t,T)P(t,T)}. \tag{1.15}$$

We purposedly used the notation $L(t,T)$ of the LIBOR rates mentioned earlier and modeled in the last chapter of the book, which are in fact simply compounded rates with the Actual/360 convention. The annually compounded spot interest rate prevailing at time t for maturity T is the constant rate $r = Y(t,T)$ for which an investment of $P(t,T)$ at time t will produce a cash flow of 1 at maturity as above, when re-investing the proceeds once a year. In other words, $Y(t,T)$ is the number r such that $(1+r)^{x(t,T)}P(t,T) = 1$ or equivalently:

$$Y(t,T) = \frac{1}{P(t,T)^{1/x(t,T)}} - 1. \tag{1.16}$$

We can now justify more rigorously the loose statement of the continuous time limit alluded to at the beginning of this section. Indeed, if for each integer $k \geq 1$ we denote by $Y^{(k)}(t,T)$ the spot interest rate compounded k times a year, then $Y^{(k)}(t,T)$ is the number r satisfying

$$\left(1 + \frac{r}{k}\right)^{kx(t,T)} P(t,T) = 1$$

or equivalently:

$$Y^{(k)}(t,T) = \frac{k}{P(t,T)^{1/kx(t,T)}} - k,$$

from which it easily follows that

$$\lim_{k\to\infty} Y^{(k)}(t,T) = -\frac{1}{x(t,T)} \log P(t,T) = r(t,T).$$

Finally, notice that we can recover (or equivalently define) the instantaneous spot rate r_t for each fixed t by

$$\begin{aligned} r_t &= \lim_{T \searrow t} r(t,T) \\ &= \lim_{T \searrow t} L(t,T) \\ &= \lim_{T \searrow t} Y(t,T) \\ &= \lim_{T \searrow t} Y^{(k)}(t,T) \end{aligned}$$

for each fixed k. We now close this subsection with a discussion of the forward rates in the context of continuous time finance. If t, T and S are dates in this order, the continuously compounded forward interest rate prevailing at time t for the time interval between T and S is the constant rate $r = f(t,T,S)$ for which an investment of $P(t,T)$ at time T will produce a cash flow of $P(t,S)$ at time S. In other words, $f(t,T,S)$ is the number r such that $e^{rx(T,S)}P(t,S) = P(t,T)$ or equivalently:

$$f(t,T,S) = -\frac{\log P(t,S) - \log P(t,T)}{x(t,T)}. \tag{1.17}$$

It is then natural to define the instantaneous forward interest rate prevailing at time t for maturity T as the number $f(t,T)$ defined as the limit:

$$f(t,T) = \lim_{x(T,S) \searrow 0} f(t,T,S)$$

from which one easily sees that

$$f(t,T) = -\frac{\partial \log P(t,T)}{\partial T} \qquad \text{or equivalently} \qquad P(t,T) = e^{-\int_t^T f(t,u)du},$$

as well as

$$r(t,T) = \frac{1}{x(t,T)} \int_t^T f(t,u)\, du.$$

1.4.3 Summary

From this point on, we shall use the notation $x = x(t,T) = T - t$, assuming that a day count convention has already been chosen, and we recap the above derivations in a set of formulae which we use throughout the book.

For each given t, we denote by $x \hookrightarrow P_t(x)$ the price of a zero coupon bond with unit nominal as a function of the time to maturity $x = T - t$, whether the latter is an integer (giving the number of years to maturity) or more generally a fraction or even a non-negative real number. With this generalization in mind, formula (1.9) gives:

$$f_t(x) = -\frac{\frac{\partial}{\partial x} P_t(x)}{P(x)} = -\frac{\partial}{\partial x} \log P_t(x).$$

Integrating both sides and taking exponentials of both sides we get the following expression for the unit zero coupon bond:

$$P_t(x) = e^{-\int_0^x f_t(s)ds}. \tag{1.18}$$

If we rewrite formulae (1.18) using the notation $x = T - t$, we get:

$$P_t(x) = e^{-xr_t(x)}$$

for the relationship between the discount unit bond and the spot rate. In terms of the forward rates it reads:

$$r_t(x) = \frac{1}{x}\int_0^x f_t(s)ds. \tag{1.19}$$

This relation can be inverted to express the forward rates as function of the spot rate:

$$f_t(x) = r_t(x) + x\frac{\partial}{\partial x}r_t(x).$$

1.5 Related Markets

Issuing a bond is the simplest economic form of borrowing, and central banks do not have exclusive rights: individuals, corporations, municipalities, counties, states, sovereigns, etc. use all forms of notes and bond issues to borrow money. For the sake of illustration, we plot in Fig. 1.5 the historical evolution of the US National Debt from 01/04/1993 through 09/01/2005.

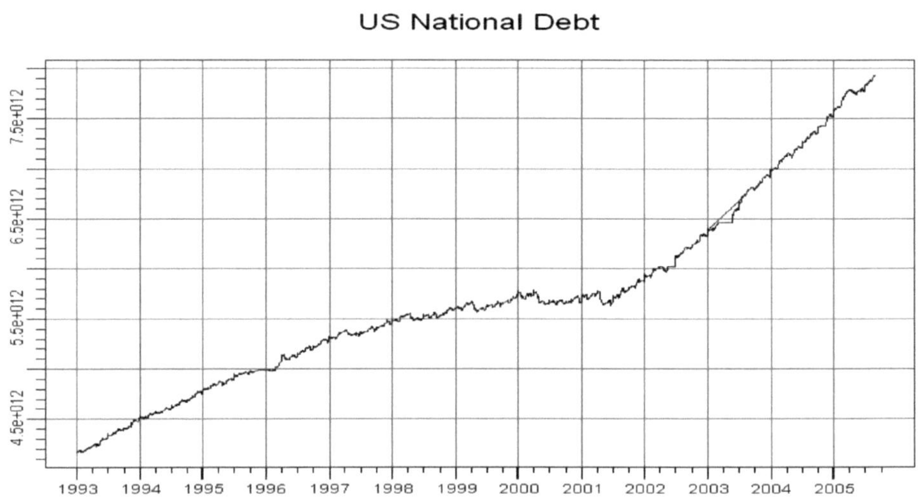

Fig. 1.5. US National Debt from 01/04/1993 through 09/01/2005. Source: Treasury Direct, `http://www.treasurydirect.gov/`.

Unfortunately, most of the important issues associated with these *loans* will not be addressed in this book. This does not mean that we regard them as unimportant. It is only a matter of time and space, and maybe taste as well. For the sake of completeness, we briefly review some of those who have been modeled for practical analysis and pricing. References are given in the Notes & Complements at the end of the chapter.

1.5.1 Municipal Bonds

Municipal bond is a generic terminology introduced to cover debt securities issued by states, cities, townships, counties, US Territories and their agencies. The interest income of these securities was exempt of federal taxes up until the Tax Reform Act of 1986. This tax advantage was one of the attractive features which made the municipal bonds (munis for short) very popular. If the interest income of all the securities issued before 1986 remains tax exempt, the situation is more complex for the securities issued after that date. The primary offerings of municipal issues are usually underwritten by specialized brokerage firms. Even though instances of default have not been plentiful, several high-profile events have given publicity to the credit risk associated with the municipal securities: we shall only mention the City of New York defaulting in 1975 on a note issue, and the highly publicized bankruptcy of Orange County in 1994. As part of the information given to the potential security buyers, municipal bond issuers hire a rating agency, sometimes even two (the most popular are S&P, Moody's and Fitch) to rate each bond issue. On the top of that, some issuers enter in a contract with an insurance company which will pay interest and principal in case of default of the issuer. So even though they are regarded as generally safe, municipal bonds carry a significant risk. As a consequence, the buyers of these securities are rewarded by a yield which is higher than the yield of a Treasury security with the same features. This difference in yield is called the yield spread over Treasury. It is expressed in basis points, and prices of municipal bonds are most often quoted by their spread over Treasury.

1.5.2 Index Linked Bonds

Index linked bonds were created in an attempt to guarantee real returns and protect cash flows from inflation. Their coupon payments and/or principal amounts are tied to a particular price index. There are four types of index linked bonds:

- *indexed-principal bonds* for which both coupons and principal are adjusted for inflation
- *indexed-coupon bonds* for which only the coupons are adjusted for inflation
- *zero coupon bonds* which pay no coupon but for which the principal is adjusted for inflation

- *indexed-annuity bonds* which pay inflation-adjusted coupons and no principal on redemption.

In the US, the most common index used is the Consumer Price Index (CPI for short) as it is the most widely used measure of inflation. It provides information about price changes in the nation's economy, and it is used as a guide to making economic decisions and formulating monetary policy. The CPI is also used as a deflator to translate prices into inflation-free dollars. The principal and the interest payments (twice a year) of Treasury Inflation-Protected Securities (TIPS for short) rise with inflation and fall with deflation. They are issued with maturities of 5, 10 and 20 years in minimal amount of $ 1000.

Index linked bonds seem to be more popular in Europe than in the US. Figure 1.6 shows how inflation-indexed Treasury securities are quoted daily in *The Wall Street Journal*.

Inflation-Indexed Treasury Securities

RATE	MAT	BID/ASKED	CHG	*YLD	ACCR PRIN
3.375	01/07	103-00/00	−3	0.969	1234
3.625	01/08	105-07/08	−4	1.259	1210
3.875	01/09	107-20/21	−7	1.459	1192
4.250	01/10	110-26/27	−7	1.604	1162
0.875	04/10	96-16/17	−7	1.677	1032
3.500	01/11	108-29/30	−8	1.715	1123
3.375	01/12	109-16/17	−11	1.759	1101
3.000	07/12	107-16/17	−13	1.811	1087
1.875	07/13	99-28/29	−13	1.888	1065
2.000	01/14	100-19/20	−13	1.917	1056
2.000	07/14	100-19/20	−13	1.922	1037
1.625	01/15	97-07/08	−14	1.951	1024
1.875	07/15	99-06/07	−15	1.963	1005
2.375	01/25	104-18/19	−25	2.084	1037
3.625	04/28	127-24/25	−35	2.074	1209
3.875	04/29	133-17/18	−38	2.066	1189
3.375	04/32	128-19/20	−41	1.989	1096

*Yield to maturity on accrued principal.

Fig. 1.6. *Wall Street Journal* inflation-indexed Treasury securities as quoted on October 15, 2005.

The first column gives the rate while the second column gives the month and year of maturity. The third column gives the bid and asked prices, while the fourth column gives the change since the quote of the last trading day. The fifth column gives the yield to maturity on the accrued principal as given in the last column.

1.5.3 Corporate Bonds and Credit Markets

Corporations raise funds in a number of ways. Short-term debts (typically less than five years) are handled via bank loans. For longer periods, commercial banks are reluctant to be the source of funds and corporations usually use bond offerings to gain access to capital. As for municipal bonds, each issue is rated by S&P and Moody's Investor Services and sometimes Fitch, and the initial rating is a determining factor in the success of the offering. These ratings are updated periodically, usually every six months after inception. They used to be the main source of information for buyers and potential buyers to quantify the credit risk associated with these bonds. For this reason, they are determining factors in the values of the bonds, and a change in rating is usually accompanied by a change in the spread over Treasury (even though it could also work the other way around). Bond with poor ratings are called non-investment grade bonds or junk bonds. Their spread over Treasury is usually relatively high, and for this reason they are also called high-yield bonds. Bond issues with the best ratings are safer; they are called investment grade bonds and their spread over Treasury is smaller.

The following gives a comparison of the yields of Treasury issues with municipal issues of comparable maturities as provided by Delphis Hanover on September 2nd, 2005, the correction for for the tax advantages was computed for a tax equivalent based on a 33% bracket.

```
Bond Yields

*Treasury Issues*

    Maturity    Coupon       Price*  Yield
    08/31/07    *4.000*      100.14  3.769
    08/15/10    *4.125*      101.07  3.855
    08/15/15    *4.250*      101.23  4.040
    02/15/31    *5.375*      116.20  4.295
*As of 4 p.m. ET. Data following decimal represent 32nds.

*Municipal Issues* (Comparable Maturities)

     AAA   Tax Equiv.   Muni/Treas   52-Week Ratio
    Yield               Yield Ratio   High     Low

     2.84      *4.24*        75.4     77.3    66.4
     3.13      *4.66*        81.1     82.8    74.1
     3.59      *5.35*        88.8     91.1    82.8
     4.39      *6.56*       102.3    106.0    93.5
```

The indenture of a corporate bond can be extremely involved. Indeed, some corporate bonds are callable (as already mentioned, some Treasury issues do have this feature too), others are convertible. Callable bonds give the issuer the option to recall the bond under some conditions. Convertible bonds give the buyer the option to convert the debt under some conditions, into a specific number of shares of the company stock. All these features make the fair pricing of the issues the more difficult. Even though they present real mathematical challenges, they will not be considered in this book.

1.5.4 Tax Issues

Tax considerations may very much change the attractiveness of some issues. In fact, tax incentives is one of the reasons corporations issue debt: not only do they need cash to refinance maturing debts or engage in capital investment, but they often want to take advantage of the tax incentives. For any given firm, finding the right balance between asset value and the level of debt is a difficult challenge, and optimal management of the capital structure of a firm in a dynamic setting is still one of the main challenges of financial economics.

Here is a more mundane example. As we already mentioned, income from coupon interest payments on Treasury notes and bonds is exempt from state income taxes. Also, interest income and growth in principal of TIPS are exempt from state and local income taxes.

Continuous compounding is a reasonable model for zero coupon bonds because they automatically reinvest the interest earnings at the rate the bond was originally bought. This feature is very attractive to some investors, except for the fact that the IRS (our friendly Internal Revenue Service) requires some bond owners to report these earned interests (though not paid). This explains in part why zero coupon bonds and STRIPS are often held by institutional investors and accounts exempt from federal income tax. They include pension funds and individual retirement accounts such as IRA and Koegh plans.

These anecdotal remarks barely touch the tip of the iceberg. Tax issues are much too intricate and technical to be discussed at the level of this book.

1.5.5 Asset Backed Securities

Once mortgage loans are made, individual mortgages are pooled together into large bundles which are securitized in the form of bond issues backed by the interest income of the mortgages. Prepayments and default risks are the main factors entering into the pricing of these securities. But the success of this market has encouraged the securitization of many other risky future incomes, ranging from catastrophic risk due to natural disasters such as earthquakes and hurricanes, weather, including temperature and rainfall, to intellectual property such as the famous example of the Bowie bond issued by the rock star, borrowing on the future cash flow expected from its rights on sales of his music. As before, the complexity of the indenture of these bonds is a challenge for the modelers trying to price these issues, and most often Monte-Carlo computations are the only tools available for pricing and risk management.

1.6 Statistical Estimation of the Term Structure

In the first part of the chapter we gave a general overview of the bond markets. We hinted at the fact that empirical data comprise real numbers corresponding to finitely many time periods and discrete times of and to maturity. Using continuous time mathematics to describe them is a modeling decision which we make for two reasons. First, we treat the time t at which the markets are observed and trades are taking place as a continuous variable. In particular, we shall use continuous discounting in all the *present value* computations. Also, we shall consider practical data as discrete (and possibly indirect) observations even though theoretical models assume continuously evolving quantities. Second, we assume that there exists a continuum of maturities T with traded bonds. In other words, we assume that the term structure of interest rates is given at least *theoretically,* by a function of the continuous variable T, considering the maturities for which we actually have quotes as discrete samples of (possibly noisy) observations. In this way, the mathematics of functional analysis can be used to define and study the models, while in order to fit the models and validate (or invalidate) them, the tools of parametric and nonparametric statistics can be brought to bear. This section is a first step in this direction. It reviews some of the statistical techniques used to infer continuous time discount, yield and forward curves from the discrete data available to the market participants. Most of the details of this section can be skipped in a first reading of the manuscript. Their *raison d'être* is the justification of the function spaces chosen as hosts of the term structure models introduced later in the book and of the types of theoretical questions addressed subsequently.

1.6.1 Yield Curve Estimation

This section reviews some of the methods of yield curve estimation used by the fixed income desks of most investment banks as well as the central banks which report to the Bank for International Settlements (BIS for short). Except for the US and Japan which use nonparametric smoothing techniques based on splines, most central banks use parametric estimation methods to infer smooth curves from the finitely many discrete values of the bond prices available each day. Parametric estimation is appropriate if the set of yield curves can be parameterized by a (small) finite number of parameters.

In Chap. 6 we will model the interest rate term structure by choosing a (possibly infinite dimensional) function space F as host of all possible forward curves. The parametric estimation procedure that we describe below is useful if the curves that occur in real life are confined to a finite dimensional (possibly nonlinear) manifold in this infinite dimensional function space. We will revisit this point in the next chapter devoted to factor models and in Chap. 6 in the context of dynamic forward rate models and the study of consistency.

The use of parametric estimation methods is partially justified by the principal components analysis discussed in Sect. 1.7 below. Indeed, we will show that the effective dimension of the space of yield curves is low, and consequently, a small number of parameters should be enough to describe the elements of this space.

Moreover, another advantage of the parametric approach is the fact that one can estimate the term structure of interest rates by choosing to estimate first the par yield curves, or the spot-rate curves, or the forward rate curves, or even the discount factor curves as functions of the maturity. Indeed, which one of these quantities is estimated first is irrelevant: once the choices of a set of curves and their parameterization are made, the parameters estimated from the observations, together with the functional form of the curves, can be used to derive estimates of the other sets of curves. We shall most often parameterize the set of forward rate curves, and derive formulae for the other curves (yields, spot rates, discount factors, etc.) by means of the relationships made explicit earlier.

On each given day, say t, one uses the available market quotes to produce a curve $x \hookrightarrow f_t(x)$ for the instantaneous forward rates as functions of the time to maturity x. For the sake of notation convenience, we shall drop the reference to the present t in most of our discussions below. The estimation of the discount factor curve would be an easy problem, if we had quotes for zero coupon bond prices of all maturities. Unfortunately, as we saw explained earlier, these instruments have maturities of less than one year: fixed income securities of longer maturities have coupons. Because we need forward curves with long maturities, the estimation procedures will be based on observations of coupon bearing bond prices and swap rates.

1.6.2 Parametric Estimation Procedures

We introduce the parametric families in use first, and we postpone the discussion of the implementation issues to later. These parametric families were introduced to capture qualitative features observed in the data.

The Nelson–Siegel Family

This family is parameterized by a four-dimensional parameter $z = (z_1, z_2, z_3, z_4)$. It is defined by:

$$f_{NS}(x, z) = z_1 + (z_2 + z_3 x)e^{-xz_4} \tag{1.20}$$

where z_4 is assumed to be strictly positive, and as a consequence, the parameter z_1, which is also assumed to be strictly positive, gives the asymptotic value of the forward rate which we will refer to as the long rate. The value $z_1 + z_2$ gives the forward rate today, i.e. the starting value of the forward curve. Since this value $f(t, 0)$ has the interpretation of the short interest rate r_t, it is also required to be positive. The remaining parameters z_3 and z_4 are responsible for the so-called *hump*. This hump does exists when $z_3 > 0$ but it is in fact a dip when $z_3 < 0$. The magnitude of this hump/dip is a function of the size of the absolute value of z_3, while z_3 and z_4 conspire to force the location along the maturity axis of this hump/dip. Once the four parameters have been estimated, formulae for the discount factor and the zero coupon yield (or spot interest rate) can be obtained by plain integration from formulae (1.18) and (1.19) respectively. We get:

$$P_{NS}(x,z) = \exp\left(-\frac{z_2 z_4 + z_3}{z_4^2} - z_1 x + \left(\frac{z_2 z_4 + z_3}{z_4^2} + \frac{z_3}{z_4^2} x\right) e^{-z_4 x}\right) \tag{1.21}$$

and

$$r_{NS}(x,z) = \frac{z_2 z_4 + z_3}{z_4^2}\frac{1}{x} + z_1 - \left(\frac{z_2 z_4 + z_3}{z_4^2}\frac{1}{x} + \frac{z_3}{z_4^2}\right) e^{-z_4 x} \tag{1.22}$$

This family used in countries such as Finland and Italy to produce yield curves.

The Svensson Family

To improve the flexibility of the curves and the fit, Svensson proposed a natural extension to the Nelson–Siegel's family by adding an extra exponential term which can produce a second hump/dip. This extra flexibility comes at the cost of two extra parameters which have to be estimated. The Svensson family is generated by mixtures of exponential functions of the Nelson–Siegel type. To be specific, the Svensson family is parameterized by a six-dimensional parameter z, and defined by:

$$f_S(x,z) = z_1 + (z_2 + z_3 x)e^{-z_4 x} + z_5 x e^{-z_6 x}. \tag{1.23}$$

As before, once the parameters are estimated, the zero coupon yield curve can be estimated by plain integration of (1.23). We get:

$$\begin{aligned} r_{NS}(x,z) = &\left(\frac{z_2 z_4 + z_3}{z_4^2} + \frac{z_5}{z_6^2}\right)\frac{1}{x} + z_1 - \left(\frac{z_2 z_4 + z_3}{z_4^2}\frac{1}{x} + \frac{z_3}{z_4^2}\right) e^{-z_4 x} \\ &+ \left(\frac{z_5}{z_6^2}\frac{1}{x} + \frac{z_5}{z_6^2}\right) e^{-z_6 x}. \end{aligned} \tag{1.24}$$

The Svensson family is used in many countries, including Canada, Germany, France and the UK.

Practical Implementation

Parametric estimation procedures are mostly used by central banks, so we rely on information published by BIS to describe the methods used by the major central banks to produce yield curves.

Description of the Available Data

On any given day t, financial data services provide quotes (most often in the form of a bid–ask interval) for a certain number of instruments with times of maturity T_j, and times to maturity $x_j = T_j - t$. These instruments are typically futures contracts on bonds and bills, swaptions, etc., and obviously, bond prices. Bid–ask spreads may be very large, a sign of illiquidity. Statistical estimation procedures should be based on the most liquid of these instruments. Moreover, small economies may not have liquid derivative markets. For these reasons, we shall concentrate on the case where only bond prices are available. For each bond, we assume that the quote includes a price, a coupon rate, a coupon frequency, interest accrued since the last coupon payments, etc., and a few other pre-computed quantities. Often the mid-point of the bid–ask interval is used as a proxy for the price.

Including information about derivative instruments in the procedure described below can be done without affecting the rationale of the estimation strategy. It merely increases the complexity of the notation and the computations, and we shall refrain from including it.

The Actual Fitting Procedure

Let us denote by B_j the bond prices available on a given day, say t. For each value of the parameter z, we denote by $B_j(z)$ the theoretical price of the bond with exactly the same indenture (i.e. with the same nominal, same maturity, same coupon rate and frequency, etc.) as B_j. Typically, this theoretical price will be obtained using formula (1.4) – corrected for accrued interest as prescribed in formula (1.7) in the case of coupon bonds – with discount factors (or equivalently zero coupon yields) given by formula (1.21) or (1.24) above; the term structure estimation procedure boils down to finding the vector z of parameters which minimizes the quadratic loss function:

$$\mathcal{L}(z) = \sum_j w_j |B_j - B_j(z)|^2 \tag{1.25}$$

where the weights w_j are chosen as a function of the duration (1.11) and the yields to maturity of the j-th bond. The dependence of the loss function upon the parameters z appears to be complex and extremely nonlinear. Were we to include other instruments in the fitting procedure, the objective function $\mathcal{L}(z)$ would include terms of the form $|I_j - I_j(z)|^2$ where I_j denotes the market quote of the instrument, and $I_j(z)$ denotes the theoretical price (present value of the discounted cash flows of the instrument) obtained from the market model with term structure determined by the value z of the parameters. In any case, fitting the parameters, i.e. finding the z's minimizing $\mathcal{L}(z)$, depends upon delicate optimization procedures which can be very unstable and computer intensive. We comment on how different central banks handle this issue in the remarks below.

Remarks.

- ◇ Many central banks do not use the full spectrum of available times to maturity. Indeed, the prices of many short term bonds are very often influenced by liquidity problems. For this reason, they are often excluded from the computation of the parameters. For example the Bank of Canada, the Bank of England and the German Bundesbank consider only bonds with a remaining time to maturity greater than three months. The French central bank also filter out the short term instruments.
- ◇ Even though it appears less general, the Nelson–Siegel family is often preferred to its Svensson relative. One reason for that is the fact that many professionals do not believe in the presence of a second bump. Moreover, being of a smaller dimension, the model is more robust and less unstable. This is especially true for countries with a relatively small number of issues. Finland is one of them. Spain and Italy are other countries using the original Nelson–Siegel family for stability reasons.
- ◇ The bid–ask spread is another form of illiquidity. Most central banks choose the mid-point of the bid–ask interval for the value of B_j. The Banque de France does just that for most of the bonds, but it also uses the last quote for some of them. Suspicious that the influence of the bid–ask spread could overwhelm the estimation procedure, the Finnish central bank uses a loss function which is equal to the sum of squares of errors where the individual errors are defined as the distance from $B_j(z)$ to the bid–ask interval (this error being obviously 0 when $B_j(z)$ is inside this interval).
- ◇ It is fair to assume that most central banks use accrued interests and clean prices to fit a curve to the bond prices. This practice is advocated in official documents of the Bank of England and the US Treasury.
- ◇ Some of the countries relying on the Svensson family fit first a Nelson–Siegel family to their data. Once this four-dimensional optimization problem is solved, they use the argument they found, together with two other values for z_5 and z_6 (often 0 and 1), as initial values for the minimization of the loss function for the Svensson family. And even then, these banks opt for the Svensson family only when the final z_5 is significantly different from 0 and z_6 is not too large! These extra steps are implemented in Belgium, Canada and France.

1.6.3 Nonparametric Estimation Procedures

We now discuss some of the nonparametric procedures used to produce forward and yield curves.

A First Estimation of the Instantaneous Forward Rate Curve

The first procedure we present was called *iterative extraction* by its inventors, but it is known *on the street* as the *bootstrapping method.* We warn the reader that this use of the word bootstrapping is more in line with the everyday use of the word bootstrapping than with the statistical terminology.

If the parametric methods described above and the nonparametric smoothing spline methods reviewed below are used by central banks, the bootstrap method which we present now is the *darling* of the investment bank fixed income desks.

As before, we present its simplest form based on the availability of bond prices only. Obvious modifications can be made to accommodate prices of other instruments whenever available.

We assume that the data at hand comprise coupon bearing bonds with maturity dates $T_1 < T_2 < \cdots < T_m$, and we search for a forward curve which is constant on the intervals $[T_j, T_{j+1})$. For the sake of simplicity we shall assume that $t = 0$. In other words, we postulate that:

$$f(0,T) = f_j \qquad \text{for} \qquad T_j < T \leq T_{j+1}$$

for a sequence $\{f_j\}_j$ to be determined recursively by calibration to the observed prices. Let us assume momentarily that $f_1, \ldots, f_j$ have already been determined and let us describe the procedure to identify f_{j+1}. If we denote by X_{j+1} the principal of the $(j+1)$-th bond, by $\{t_{j+1,i}\}_i$ the sequence of coupon payment times, and by $C_{j+1,i} = c_{j+1}/n_y$ the corresponding payment amounts (recall that we use the notation c_j for the annual coupon rate, and n_y for the number of coupon payments per year), then its price at time $t = 0$ can be obtained by discounting all the future cash flows associated with this bond:

$$\begin{aligned} B_{j+1} = &\sum_{t_{j+1,i} \leq T_j} P(0,t_i) \frac{c_{j+1} X_{j+1}}{n_y} \\ &+ P(0,T_j) \left(\sum_{T_j < t_{j+1,i} \leq T_{j+1}} e^{-(t_{j+1,i}-T_j)f_{j+1}} \frac{c_{j+1} X_{j+1}}{n_y} + e^{-(t_{j+1,i}-T_j)f_{j+1}} \right). \end{aligned} \tag{1.26}$$

Notice that all the discount factors appearing in this formula are known since, for $T_k \leq t < T_{k+1}$ we have:

$$P(0,t) = \exp\left[\sum_{h=1}^{k} (T_h - T_{h-1}) f_h + (t - T_k) f_{k+1} \right]$$

and all the forward rates are known if $k < j$. Consequently, rewriting (1.26) as:

$$\begin{aligned} &\frac{B_{j+1} - \sum_{t_{j+1,i} \leq T_j} P(0,t_i) \frac{c_{j+1} X_{j+1}}{n_y}}{P(0,T_j)} \\ &\qquad = \frac{c_{j+1} X_{j+1}}{n_y} \sum_{T_j < t_{j+1,i} \leq T_{j+1}} e^{-(t_{j+1,i}-T_j)f_{j+1}} + e^{-(t_{j+1,i}-T_j)f_{j+1}} \end{aligned} \tag{1.27}$$

and since the left-hand side can be computed and the unknown forward rate f_{j+1} appears only in the right-hand side, this equation can be used to determine f_{j+1} from the previously evaluated values f_k for $k \leq j$.

Remark 1.1. Obviously, the forward curve produced by the bootstrapping method is discontinuous, since by construction, it jumps at all the input maturities. These jumps are the source of an artificial volatility which is only due to the method of estimation of the forward curve. This is the main shortcoming of this method of estimation. Several remedies have been proposed to alleviate this problem. The simplest one is to increase artificially the number of maturity dates T_j to interpolate between the observed bond (or swap) prices. Another proposal is to add a smoothness penalty which will force the estimated curve to avoid jumps. This last method is in the spirit of the smoothing spline estimation method which we present now.

Smoothing Splines and the US and Japan Forward Curves

As already stated in Sect. 1.6.1, the yield and forward curves published by the US Federal Reserve and the Bank of Japan are computed using smoothing splines. These curves, which relate the yield on a security to its time to maturity are based on the closing market bid yields on actively traded Treasury securities in the over-the-counter market. The general strategy used to produce them goes as follows. On any given day, the instantaneous forward rate curve is the function $x \hookrightarrow \varphi(x)$ which minimizes the objective function:

$$\mathcal{L}_{JUS}(\varphi) = \sum_{j=1}^{n} w_j |B_j - B_j(\varphi)|^2 + \lambda \int |\varphi''(x)|^2 \, dx \tag{1.28}$$

where $\lambda > 0$ is a parameter called the *smoothing parameter*, where $\varphi''(x)$ stands for the second derivative of $\varphi(x)$, where the B_j's are the prices for on-the-run Treasury bonds and notes available on day t, or proxies, and $B_j(\varphi)$ denotes the price one would get pricing these bonds and notes from the forward curve given by φ. A security is said to be on-the-run if it is the most recently issued US Treasury bond or note of a particular maturity. Recall also that the w_j's represent weights chosen as functions of the durations of the bonds, and the theoretical prices are computed according to the theory presented in Sect. 1.2.

Choosing a forward curve $x \hookrightarrow \varphi(x)$ which minimizes the objective function $\mathcal{L}_{JUS}$ guarantees that the forward curve is smooth because of the presence of the second term in the definition of $\mathcal{L}_{JUS}$, and at the same time it ensures a fit to the price data because of the presence of the first term. The balance between these two contributions to the objective function is controlled by the smoothing parameter λ. Obviously, large values of λ will produce very smooth curves while smaller values of λ will lead to rough curves trying to reproduce perfectly the prices observed on the market.

Standard results from nonparametric statistics show that the argument φ_{JUS} of the minimization of the objective function (1.28) is in fact a cubic spline. This fact justifies the terminology of smoothing spline for this nonparametric method of construction of forward curves.

1.7 Principal Component Analysis

This book is devoted to the analysis of mathematical models describing changes in the term structure of interest rates from one day to the next. The statistical procedures presented in the previous sections give tools to construct the term structure, say the forward curve, on a given day like today. Mathematical models for the dynamics of this curve depend on the way the curve is coded. Using parametric statistics for the estimation of the current forward curve suggests that a finite dimensional manifold can accurately describe the set of forward curves, and hence that a finite dimensional parameter z can capture the entire curve. If this manifold does not change from day to day, the dynamics of the term structure of interest rates are given by the dynamics of the characteristic parameter z, the latter being most often modeled by a finite dimensional diffusion process. On the other hand, using nonparametric statistics for the estimation of the forward curve suggests that the manifold of relevant forward curves is infinite dimensional, and that the dynamics of the term structure of interest rates need to be given by a stochastic process of an infinite dimensional nature. Both points of view are developed in the book.

In this section we look at the principal components analysis (PCA for short) of real interest rate data. In doing so, we hope to find the effective dimension of the space of yield curves. In some rather limited way, the *raison d'être* for this section is the justification of the finite-factor models used for the term structure of interest rates which we introduce in the following chapter, and to which we come back over and over throughout the book.

1.7.1 Principal Components of a Random Vector

Let X be an m-dimensional random vector with mean $\mathbb{E}\{X\} = \mu$ and covariance matrix $\mathbb{E}\{(X-\mu)\otimes(X-\mu)\} = Q$, where $u\otimes v$ means that $m\times m$ matrix whose entries are given by $(u\otimes v)_{ij} = u_i v_j$. So if u is viewed as an $m\times 1$ column vector, and the transpose ${}^t v$ as a $1\times m$ row vector, then the $m\times m$ matrix Q can be viewed as $Q = u\,{}^t v$. In operator form, the *tensor product* $u\otimes v$ can be characterized by the fact that it is the only matrix satisfying $(u\otimes v)w = u\langle v, w\rangle$ for all vectors $w\in\mathbb{R}^m$. Since Q is symmetric and positive, its spectral decomposition reads

$$Q = \sum_{i=1}^{m} \lambda_i^2 v_i\otimes v_i$$

where $\lambda_1^2 \geq \lambda_2^2 \geq \cdots \geq \lambda_m^2 \geq 0$ are the eigenvalues of Q arranged in decreasing order (possibly repeated according to their multiplicities), and $\{v_i\}_{i=1,\ldots,m}$ is an orthonormal basis of corresponding eigenvectors. Note that $\sum_{i=1}^m \lambda_i^2 = \mathbb{E}\{\|X-\mu\|^2\}$.

The random vector X can be decomposed as

$$X = \mu + \sum_{i=1}^{m} \lambda_i v_i \xi_i$$

where the scalar random variables $\xi_1, \ldots, \xi_m$ have mean zero, unit variance, and are uncorrelated in the sense that $\mathbb{E}\{\xi_j \xi_j\} = 0$ whenever $i \neq j$. Furthermore, if the random vector X is Gaussian, then the scalar random variables $\xi_1, \ldots, \xi_m$ are independent and Gaussian $N(0,1)$ random variables. That is, every realization of the random vector X can be decomposed into a linear combination of uncorrelated and orthogonal random vectors pointing in fixed directions, and in the Gaussian case, these vectors are independent. This decomposition of a random vector in the form of a series expansion will be generalized in Chap. 3 to the case of infinite dimensional Gaussian random vectors, Gaussian processes and random fields.

It is often the case that the random vector X we are studying is a model for a m-dimensional quantity that occurs in an application. If the dimension m is very high, it is difficult to have good intuition about the behavior of this quantity. However, if the eigenvalues λ_j of the covariance matrix are very close to zero for $j = d+1, \ldots, m$ for some number d which is much smaller than m, we say that the effective dimension of X is d. Indeed, the vector X can be accurately approximated by

$$X \approx \mu + \sum_{i=1}^{d} \lambda_i v_i \xi_i.$$

That is, although X takes values in an m-dimensional space, it effectively lives in the shift by μ of the d-dimensional linear subspace spanned by the first d eigenvectors $v_1, v_2, \ldots, v_d$.

We will see that this is precisely the situation we face when we study the forward rate curve. In that case, the dimension m corresponds to the number of times-to-maturity dates for which we have data. In the models studied in Chap. 6 the forward rate curve is a random element of an infinite dimensional space F. However, the eigenvalues of the covariance matrix computed with real forward rates decay very quickly. We will see that the effective dimension of the space of forward rate curves is about three or four.

1.7.2 Multivariate Data PCA

The theoretical discussion of the previous subsection underpins the best-known dimension reduction technique in multivariate data analysis. It goes under the name of principal component analysis (PCA).

In practice, data come in the form of a large matrix $\mathbf{x} = [x_{ij}]$ and we assume that each row $x^{(i)} = [x_{i1}, \dots, x_{im}]$ is a sample realization of an m-dimensional random vector $X^{(i)}$ and the data manipulations encompassed by PCA are based on the following theoretical results.

If $X^{(1)}, X^{(2)} \dots$ form a sequence of random vectors, each with the same law in $\mathbb{R}^m$, then the strong law of large numbers tells us that the *estimators*

$$\mu_N = \frac{1}{N} \sum_{i=1}^{N} X^{(i)}$$

and

$$Q_N = \frac{1}{N-1} \sum_{i=1}^{N} (X^{(i)} - \mu_N) \otimes (X^{(i)} - \mu_N)$$

converge as $N \to \infty$ almost surely to the mean vector μ and covariance matrix Q of any random vector with the same law as all the $X^{(i)}$. This theoretical result holds provided that the random vectors $X^{(1)}, X^{(2)} \dots$ are independent and identically distributed and satisfy a moment condition.

In practice, we use the data samples contained in the data matrix $\mathbf{x}$ to compute empirical estimates $\hat{\mu}_N$ and $\hat{Q}_N$ for the estimators μ_N and Q_N. These empirical estimates are usually chosen to be:

$$\hat{\mu}_N = \frac{1}{N} \sum_{i=1}^{N} x^{(i)}$$

and

$$\hat{Q}_N = \frac{1}{N-1} \sum_{i=1}^{N} (x^{(i)} - \hat{\mu}_N) \otimes (x^{(i)} - \hat{\mu}_N)$$

as the law of large numbers guarantees that, should the sample size N be large enough, these empirical values $\hat{\mu}_N$ and $\hat{Q}_N$ would be close to the desired values μ and Q of the common mean vector and variance/covariance matrix.

For our applications, however, the row indices i label the dates t for which we have historical data for the vector of quotes used to perform the analysis, and it is unlikely that the interest rates are independent from day to day. In order to overcome this problem, there are several ways to proceed. One way is to assume that the interest rate time series is stationary. Indeed, under the assumption that the sequence $X^{(1)}, X^{(2)} \dots$ is stationary, the estimators μ_N and Q_N still converge to the true mean vector and covariance matrix respectively, even though the rate of convergence may not be the same. So in what follows, we will implicitly assume that the data row vectors obtained from interest rate quotes are samples from a stationary sequence of random vectors. So in order to use the interpretation of the results of a PCA, we must remember to check that the models we develop have this stationarity property. In Chap. 7 we carry out the analysis of the stationarity for the linear Gaussian HJM model.

When the stationarity of the time series of row vectors of the data matrix $\mathbf{x}$ is in doubt, the standard practice is to *differentiate* the series and to compute the principal components of the daily *increments* of the interest rate instrument quotes. In this way, we subtract out most of the dependence of day n's interest rate term structure on that of day $n-1$'s. We do not need to take this extra step here in the examples presented in the next subsections, but this form of the principal component analysis of the increments of interest rate data can be found, for instance, in the paper of Bouchaud, Cont, El Karoui, Potters, and Sagna [23]. The results of this PCA are broadly similar to the results for the full time series. In particular, they found that a few eigenvectors explain most of the variance of the daily increments, and their shapes are of the level, slope, and curvature variety described below.

1.7.3 PCA of the Yield Curve

For the purpose of this first example, we use data on the US yield curve as provided by the US Treasury. As we are about to explain, these data have been the object of manipulations, typically interpolation and extrapolation to guarantee that on any given day one has yield quotes for the same times to maturity every day. These rates are commonly referred to as "Constant Maturity Treasury" rates, or CMT. Yields are extracted by the Treasury from the daily yield curve computed in the way described in Sect. 1.6.3. The CMT yield values are read from the yield curve at fixed maturities, currently 1, 3 and 6 months, and 1, 2, 3, 5, 7, 10, 20 and 30 years.

The extreme maturities create some mathematical challenges. Indeed, short term maturities are not always available, and the retirement of the long bond in 2001 made it difficult to have reliable extrapolated values for the constant maturity 30 years. So, we divided the data into two disjoint subsets, and report the numerical results for each of them separately.

We first consider the period ranging from 10/1/1993 to 7/31/2001 and for this period we considered all the maturities except the first one of 1 month. Our data matrix has 1961 rows, one for each of the trading days in the time span we consider, and 10 columns, one for each of the constant maturities. In other words, the columns contain the yields on the US Treasuries for times to maturity

$$x = 0.25, 0.5, 1, 2, 3, 5, 7, 10, 20, 30 \qquad \text{years.}$$

Figure 1.7 gives the proportions of the variation explained by the various components. The first three eigenvectors of the covariance matrix (the so-called loadings) explain 99.7% of the total variation in the data. This suggests strongly that, if we are not misled by a smoothing artifact produced by the pre-processing of the raw data, the effective dimension of the space of yield curves could be three. In other words, any of the yield curves from this period can be approximated by a linear combination of the first three loadings, the relative error being very small. Figure 1.8 gives the plots of the first four loadings.

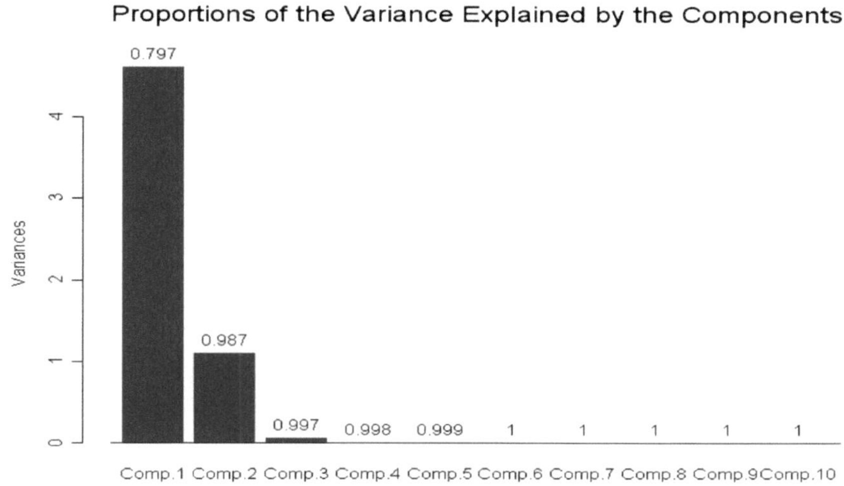

Fig. 1.7. Proportions of the variance explained by the components of the PCA of the daily US Treasury yields.

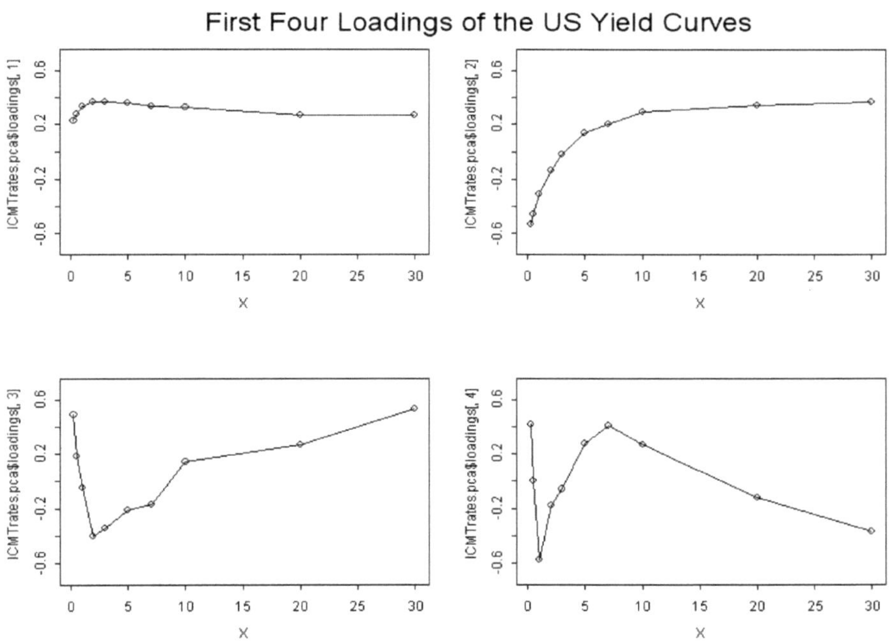

Fig. 1.8. From left to right and top to bottom, sequential plots of the first four US Treasury yield loadings for the period 10/1/1993 to 7/31/2001. We changed the scale of the horizontal axis to reflect the actual times to maturity.

The first loading is essentially flat, so a component on this loading will essentially represent the average yield over the maturities. Because of the monotone and increasing nature of the second loading, the second component measures the upward trend (if the component is positive and the downward trend otherwise) in the yield. The shape of the third loading suggests that the third component captures the curvature of the yield curve. Finally, the shape of the fourth loading does not seem to have an obvious interpretation. It is mostly noise (remember that most of the variations in the yield curve are explained by the first three components). These features are very typical, and they should be expected in most PCA of the term structure of interest rates.

The fact that the first three components capture so much of the features of the yield curve may seem strange when compared to the fact that some estimation methods which we discussed use parametric families with more than three parameters! There is no contradiction there.

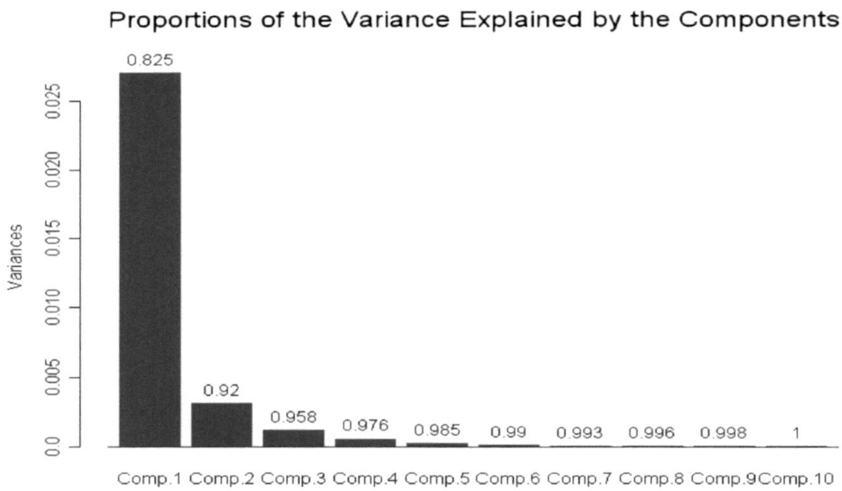

Fig. 1.9. Proportions of the variance explained by the components of the PCA of the daily changes in US Treasury yields over the period 8/1/2001 to 8/12/2005.

We now consider the data provided by the Treasury for the period ranging from 8/1/2001 to 8/12/2005. This period spans 1008 trading days, and we still use 10 maturities as we need to drop the long bond yield while we gain the one-month yield. The results are almost identical to the results reported above. So instead of presenting them, we use this new data set to illustrate the discussion of the stationarity of Sect. 1.7.2. We first compute the daily changes in yield for the 10 maturities in question. The number of rows of our matrix is now 1007, while we still have the same constant times to maturity:

$$x = 1/12, 1/4, 1/2, , 1, 2, 3, 5, 7, 10, 20 \qquad \text{years.}$$

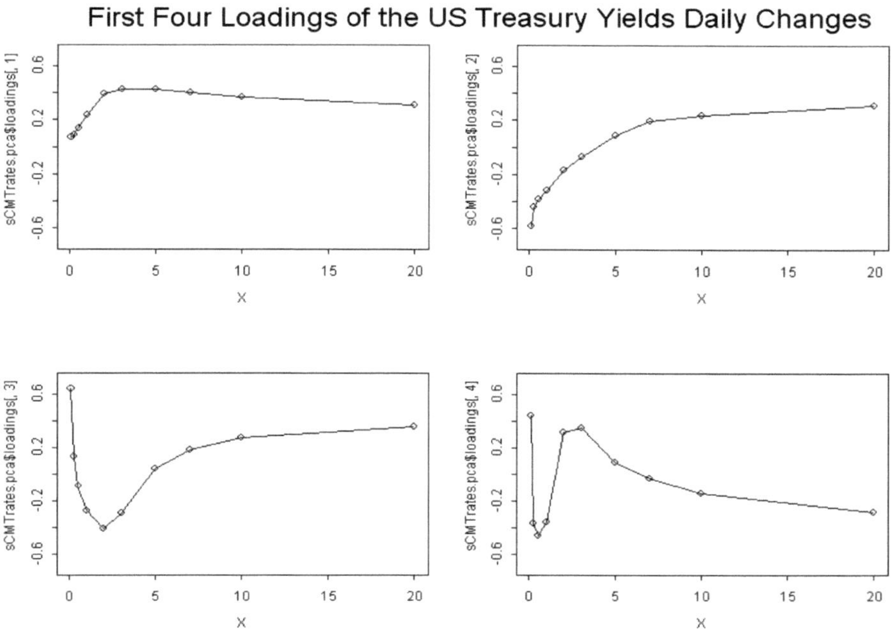

Fig. 1.10. From left to right and top to bottom, sequential plots of the first four US Treasury yield loadings of the PCA of the daily changes in US Treasury yields over the period 8/1/2001 to 8/12/2005.

Figure 1.9 gives the proportions of the variation explained by the various components. The proportion of the fluctuation explained by the first three eigenvectors is slightly less as it is now 95.8% of the total variation in the data. Figure 1.10 gives the plots of the first four loadings. Obviously the interpretation remains the same: the change in yield curve from one day to the next is composed of a linear superposition of a horizontal shift, a tilt and a curvature components.

1.7.4 PCA of the Swap Rate Curve

Figure 1.11 gives the proportions of the variation explained by the various components while Fig. 1.12 gives the plots of the first four eigenvectors.

Our second application of principal component analysis concerns the swap rate curves described earlier in Sect. 1.3.3. As before, we denote by m the dimension of the vectors. We use data downloaded from `Data Stream`. Again, it is quite likely that the raw data have been processed, but we are not quite sure what kind of manipulation is performed by `Data Stream` so for the purpose of this illustration, we shall ignore the possible effects of the pre-processing of the data. In this example, the day t labels the rows of the data matrix. The latter has $M = 14$ columns, containing constant maturity swap rates with times to

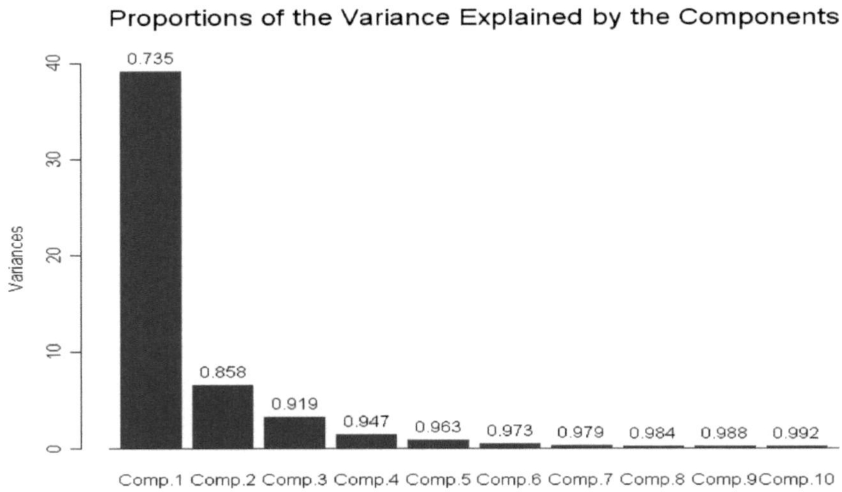

Fig. 1.11. Proportions of the variance explained by the components of the PCA of the daily changes in the swap rates for the period from August 1998 to October 2005.

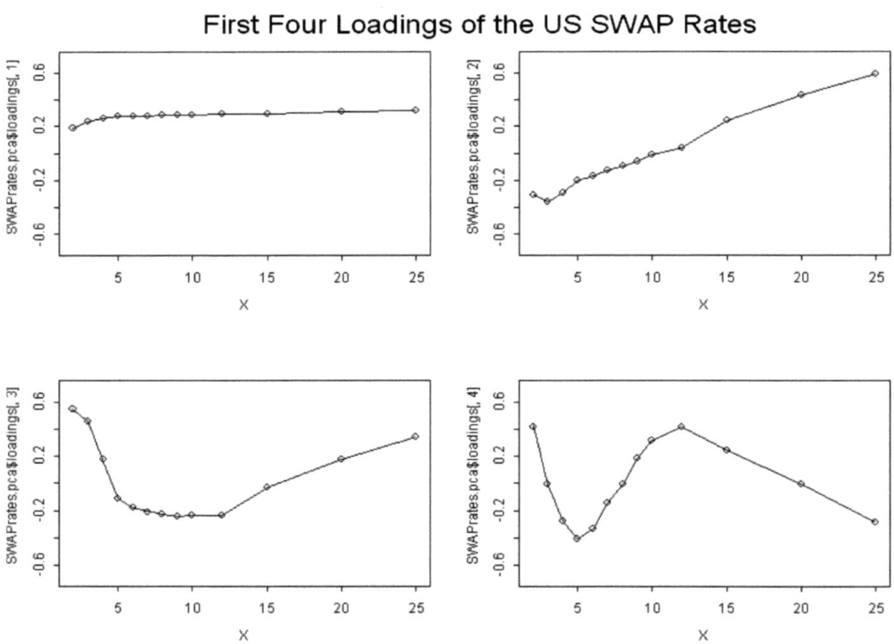

Fig. 1.12. From left to right and top to bottom, sequential plots of the eigenvectors (loadings) corresponding to the four largest eigenvalues. As before, we changed the scale of the horizontal axis to reflect the actual times to maturity.

maturity $x = T - t$ which have the values $2, 3, 4, 5, 6, 7, 8, 9, 10, 12, 15, 20, 25$ and 30 years. We collected these data for each day t of the period from August 7, 1998 to October 7, 2005, and we computed the covariance matrix of the daily changes after we drop the last column because of the possible artifacts created by the retirement of the 30 year bond. We rearranged these numerical values in a matrix $R = [r_{i,j}]_{i=1,\cdots,N,\ j=1,\cdots,M}$. Here, the index j stands for the time to maturity while the index i codes the day the curve is observed.

The interpretation of these four loadings is clear from Fig. 1.12. The first component of the daily change in swap rate is a horizontal shift of the overall level of the swap rate, while the second component measures a trend or tilt in the rate, and the third component tries to capture a curvature effect.

Since such an overwhelming proportion of the variation is explained by one single component, it is often recommended to remove the effect of this component from the data (here, that would amount to subtracting the overall mean rate level) and to perform the PCA on the transformed data (here, the fluctuations around the mean rate level).

Notes & Complements

For more information on the mechanics of the fixed income markets, and for details on mortgage-backed securities, the reader is referred to Fabozzi's encyclopedic handbooks [56] and [57]. Rebonato's books [115] and [114] present the practitioner point of view on the complete zoology of the fixed income derivatives which are not discussed in this book, together with market data issues and the LIBOR markets barely touched upon in the last chapter of the book. The reader interested in a more mathematical treatment is referred to the textbook of Brigo and Mercurio [25].

The highly publicized defaults of counties (such as the bankruptcy of Orange County in 1994), of sovereigns (like Russia defaulting on its bonds in 1998) and the ensuing ripple effects on worldwide markets have brought the issue of credit risk to the forefront. For the last ten years, credit risk has been a steadily growing component of the fixed income desk of most investment banks. The market of credit derivatives is growing at an exponential pace, and the rate of increase in appetite for these products seems to be limitless. Hedge funds and proprietary trading desks of major banks have become heavy users and theoretical research is enjoying a piggyback ride. Unfortunately, because of time and space limitations, we cannot address these issues in this book. The reader interested in learning about credit risk is referred to the many textbooks which recently appeared on this subject. We shall give references to the four books which we consider as the most mathematical of the lot: Bielecki and Rutkowski [7], Schönbucher [122], Duffie and Singleton [51] and Lando [99].

The Nelson–Siegel and Svensson families of parameterized yield curves have enjoyed much popularity among central bankers. It should be noted, though, that these families were introduced to solve a specific *static* problem: to find a curve described by a few parameters that fit a single day's bond prices reasonably well. One may ask, then, how do these parameters vary from day to day? As we will discuss in the next chapter, a reasonable *dynamic* model of the term structure should reflect the economic principle that the bond market is free of arbitrage opportunities. Now, consider a hypothetical dynamic model of the term structure that has the property that each day the forward rate curve is in the Nelson–Siegel family. Essentially, it turns out that such a model necessarily admits arbitrage! This surprising fact was proven by Filipović, see [62]. We will return to these issues of arbitrage and consistency in Chap. 6.

Principal component analysis is a classical tool of data analysis which originated in the statistical theory of linear models, and which has been used in many other fields in which practitioners have to handle data sets of high dimensions. The use of this technique for the analysis of the term structure of interest rates seems to have originated in the work of Litterman and Scheinkman [101]. Other sources include Rebonato's book [115] or the paper by Bouchaud, Sagna, El Karoui and Cont [23]. The numerical results reported in this chapter are updated versions of those given in Chap. 2 of Carmona's textbook on statistical analysis of financial data [32] where more explanations about the mathematical underpinnings of the PCA algorithm can be found.

2

Term Structure Factor Models

This chapter gives a first introduction to stochastic models for the term structure of interest rates. We try to give some perspective on the historical development of these models. Despite the fact that the main thrust of the book is the analysis of genuinely infinite dimensional models, we first review the more classical approach, modeling the term structure of interest rates by means of a finite number of stochastic factors. The latter offer intuitive ways to impose a mathematical structure on the short interest rate. This approach has been (and still is) very popular because of its tractability and the wealth of results leading to easy implementations. The Markov property present in most of these models leads to partial differential equation formulations and to reasonable expressions for derivative prices. Also, the class of affine models leads to closed-form formulae for many instruments and derivatives. We frame the presentation in a way that will ease the transition to the models discussed in the last part of the book. The chapter closes with a discussion of the HJM framework. This framework is central to later chapters of the book as we begin to see the infinite dimensional character of the term structure.

2.1 Factor Models for the Term Structure

The goal of this chapter is to discuss a time-honored method for pricing interest rate securities in a continuous time economy. This is not the most general framework for pricing interest rate securities, but it will suffice in showing the role of the stochastic building blocks, especially in the no-arbitrage arguments.

We assume that at each time t, the price $P(t,T)$ of a zero coupon bond with maturity T and nominal value \$1 is a random variable defined on a probability space $(\Omega, \mathcal{F}, \mathbb{P})$. The maturity date T satisfies $0 < t \leq T < +\infty$. We shall sometimes use the notation $\mathcal{I}$ for the set of couples (t,T) with $0 < t \leq T < +\infty$, where we allow t and T to wander all the way to ∞ for mathematical convenience. Being a random variable will not be sufficient to do analysis. We usually add the more restrictive assumption that $T \hookrightarrow P(t,T)$ is almost surely smooth. In this case, the instantaneous forward rate $f(t,T)$ at time t for maturity T is well-defined and given by the formula

$$f(t,T) = -\frac{\partial}{\partial T} \log P(t,T).$$

Recall the short interest rate r_t is given by the formula $r_t = f(t,t)$. We assume that the probability space $(\Omega, \mathcal{F}, \mathbb{P})$ is equipped with a filtration $\{\mathcal{F}_t\}_{t \geq 0}$ which defines history in the sense that the elements of $\mathcal{F}_t$ are the events prior to time t, and we assume that for each fixed $T > 0$, the prices $\{P(t,T);\ 0 \leq t \leq T\}$ of the zero coupon bonds with maturity T form a stochastic process adapted to this filtration.

We say that the term structure of interest rates is given by a factor model if there exists an Itô process $\{Z_t\}_{t \geq 0}$ taking values in an open subset $D \subset \mathbb{R}^k$ and if the prices of the zero coupon bonds are given by an equation of the form

$$P(t,T) = P^{(Z)}(t,T,Z_t) \tag{2.1}$$

for some real-valued (deterministic) function $(t,T,z) \hookrightarrow P^{(Z)}(t,T,z)$ on $\mathcal{I} \times D$. The components of the vector Z_t are called the factors of the model. In most of the applications considered here, D will be the whole Euclidean space $\mathbb{R}^k$ or a halfspace. Taking logarithmic derivatives of both sides of Eq. (2.1) shows that a factor model can be defined by assuming that the forward rates are of the form:

$$f(t,T) = f^{(Z)}(t,T,Z_t) \tag{2.2}$$

for some real-valued (deterministic) function $(t,T,z) \hookrightarrow f^{(Z)}(t,T,z)$ on $\mathcal{I} \times D$. Factor models are often defined from Eq. (2.2) rather than from (2.1). Obviously we have:

$$P^{(Z)}(t,T,z) = e^{-\int_t^T f^{(Z)}(t,u,z)du}$$

and the short interest rate r_t is given by $r_t = f^{(Z)}(t,t,Z_t)$, and the model is arbitrage free (i.e. $\mathbb{P}$ is a local martingale measure) if and only if

$$\left\{ e^{-\int_t^T f^{(Z)}(u,u,Z_u)du} P^{(Z)}(t,T,Z_t) \right\}_{0 \leq t \leq T} \tag{2.3}$$

is a local martingale for all $T > 0$. See the Appendices at the end of the chapter for more information on this issue.

The assumption that $\{Z_t\}_{t \geq 0}$ is an Itô process is usually strengthened by assuming that it is in fact an Itô diffusion in the sense that it satisfies a stochastic differential equation of the form

$$dZ_t = \mu^{(Z)}(t,Z_t)dt + \sigma^{(Z)}(t,Z_t)dW_t \tag{2.4}$$

where the drift is given by a smooth (deterministic) $\mathbb{R}^k$-valued function $(t,z) \hookrightarrow \mu^{(Z)}(t,z)$ on $[0,\infty) \times D$, the diffusion part is given by a smooth (deterministic) function $(t,z) \hookrightarrow \sigma^{(Z)}(t,z)$ on $[0,\infty) \times D$ with values in the space of $k \times k$ matrices, and $\{W_t\}_{t\geq 0}$ is a standard k-dimensional Wiener process. The factors Z_t are intended to encapsulate the part of the state of the economy at time t which is relevant to the term structure of interest rates, and the functional form (2.2) quantifies this dependence mathematically. In order to emphasize the fact that the components of Z_t are the only factors affecting the interest rates dynamics, we assume that the history is given by the filtration $\{\mathcal{F}_t^{(Z)}\}_{t\geq 0}$ generated by the factors. In fact, in order to avoid pathologies which could be created by degeneracies of the diffusion equation (2.4), we usually assume that the matrices are non-singular, and as a consequence, this history is as well given by the filtration generated by the Wiener process and

$$\mathcal{F}_t = \mathcal{F}_t^{(Z)} = \mathcal{F}_t^{(W)}, \qquad\qquad t \geq 0.$$

Applying Itô's formula to (2.2) given (2.4) implies that $\{P(t,T)\}_{0\leq t\leq T}$ is an Itô process for each fixed $T > 0$, and its stochastic differential is of the form

$$dP(t,T) = \mu^{(P)}(t,Z_t)dt + \sigma^{(P)}(t,Z_t)dW_t \tag{2.5}$$

for some deterministic functions $(t,z) \hookrightarrow \mu^{(P)}(t,z)$ and $(t,z) \hookrightarrow \sigma^{(P)}(t,z)$. Notice that since:

$$r_t = \lim_{T\searrow t} -\frac{1}{T-t}\log P(t,T) = \lim_{T\searrow t} -\frac{1}{T-t}\log P^{(Z)}(t,T,Z_t) \tag{2.6}$$

it is conceivable that the short interest rate process $\{r_t\}_{t\geq 0}$ is also an Itô process on the same filtration generated by the k-dimensional Wiener process $\{W_t\}_{t\geq 0}$.

As it stands now, there is a lot of flexibility built into the possible factor models. We will concentrate our attention on two simple cases which are very popular in practice: the affine models and the short rate models. The affine models are those for which one makes the simplifying assumption that the function $z \hookrightarrow f^{(Z)}(t,T,z)$ is affine in the sense that that

$$f^{(Z)}(t,T,z) = A(t,T) + B(t,T)z.$$

We will see that such models are very tractable. The short rate models, on the other hand, are those factor models for which the economic factor Z_t is taken to be one-dimensional, and in fact, equal to the short interest rate r_t. Notice that these classes are not disjoint. Indeed, the affine short rate models are particularly nice since many explicit calculations are possible.

2.2 Affine Models

We say that a factor model as above is *exponential-affine* if the function $P^{(Z)}$ is of the form

$$P^{(Z)}(t,T,z) = e^{A(t,T)+B(t,T)z} \tag{2.7}$$

for some smooth functions A and B on $\mathcal{I}$. Here for each couple $(t,T) \in \mathcal{I}$, $A(t,T)$ is a scalar while $B(t,T)$ is a k-dimensional vector. Notice that we necessarily have:

$$A(t,t) = 0 \qquad \text{and} \qquad B(t,t) = 0$$

because $P^{(Z)}(t,t,z) = 1$ for all $z \in D$. The affine property can equivalently be defined in terms of the instantaneous forward rates. It reads:

$$f^{(Z)}(t,T,z) = \tilde{A}(t,T) + \tilde{B}(t,T)z \tag{2.8}$$

where the functions $\tilde{A}(t,T)$ and $\tilde{B}(t,T)$ are merely the negative partial derivatives of $A(t,T)$ and $B(t,T)$ of (2.7) with respect to the time of maturity T. The terminology *affine* has its origin in formula (2.8). Expanding the latter we see that the instantaneous forward rates are given by:

$$f(t,T) = \tilde{A}(t,T) + \tilde{B}_1(t,T)Z_t^{(1)} + \cdots + B_k(t,T)Z_t^{(k)}$$

which show that the scalar functions $\tilde{B}_1(t,T), \ldots, \tilde{B}_k(t,T)$ can be interpreted as the sensitivities of the forward rate to the factor loadings.

Moreover if, in line with our earlier discussion, we assume the existence of a deterministic function $(t,z) \hookrightarrow r^{(Z)}(t,z)$ satisfying

$$r^{(Z)}(t,z) = \lim_{T \searrow t} -\frac{1}{T-t} \log P^{(Z)}(t,T,z), \qquad (t,z) \in [0,\infty) \times D, \tag{2.9}$$

then it follows that $r^{(Z)}$ is affine in the sense that

$$r^{(Z)}(t,z) = a(t) + b(t)z, \qquad (t,z) \in [0,\infty) \times D, \tag{2.10}$$

with $a(t) = \frac{\partial}{\partial T}A(t,T)|_{T=t} = -\tilde{A}(t,t)$ and $b(t) = \frac{\partial}{\partial T}B(t,T)|_{T=t} = -\tilde{B}(t,t)$.

If we restrict ourselves to the homogeneous case for the sake of simplicity,

$$\mu^{(Z)}(t,z) = \mu^{(Z)}(z) \quad \text{and} \quad \sigma^{(Z)}(t,z) = \sigma^{(Z)}(z),$$

and the functions $P^{(Z)}$, and hence $f^{(Z)}$, depend only on the time to maturity $x = T-t$, i.e. $P^{(Z)}(t,T,z) = P^{(Z)}(x,z)$ and $f^{(Z)}(t,T,z) = f^{(Z)}(x,z)$. In this case, the no-arbitrage condition reads:

$$\partial_x \tilde{A}(x) + \partial_x \tilde{B}(x) \cdot z = \mu^{(Z)}(z) \cdot \tilde{B}(z) - \frac{1}{2}\partial_x B(x)^* a(x) B(x).$$

See the Appendices at the end of the chapter for a discussion of the general case. Recall that with the notation of the homogeneous case, we have $\tilde{A}(x) = -A'(x)$ and $\tilde{B}(x) = -B'(x)$. From this it easily follows that if the functions

$$B_1, \ldots, B_k, B_1^2, B_1 B_2, \ldots, B_k^2$$

are linearly independent, then the drift and the diffusion coefficients $\mu^{(Z)}(z)$ and $a^{(Z)}(z)$ are affine functions of z in the sense that

$$\begin{cases} \mu^{(Z)}(z) = \mu^{(Z)} + \overline{\mu}^{(Z)} z \\ a^{(Z)}(z) = \alpha^{(Z)} + \sum_{i=1}^{k} \alpha_k^{(Z)} z \end{cases} \tag{2.11}$$

for some constant k-vector $\mu^{(Z)}$ and $k \times k$-matrices $\overline{\mu}^{(Z)}$, $\alpha^{(Z)}$ and $\alpha_k^{(Z)}$. Plugging formulae (2.11) into the no-arbitrage condition gives the following system of ordinary differential equations:

$$\begin{cases} \partial_x A(x) = -\tilde{A}(0) + \mu^{(Z)} \cdot B(x) - \frac{1}{2} B(x)^* \alpha^{(Z)}(x) B(x) \\ \partial_x B(x) = -\tilde{B}(0) + \overline{\mu}^{(Z)} B(x) - \frac{1}{2} B(x)^* \alpha_{\cdot}^{(Z)}(x) B(x) \end{cases} \tag{2.12}$$

with the initial conditions $A(0) = 0$ and $B_i(0) = 0$, for $i = 1, \ldots, k$. Note that the second equation is a k-dimensional Riccati equation in $B(x)$. Once solved, one can plug the resulting $B(x)$ in the first equation and get $A(x)$ by a ordinary integration.

Part of the above discussion can be recast in the following result which we state without making fully explicit the mild technical conditions under which it holds.

Proposition 2.1. *The term structure factor model is exponential affine if and only if the drift $\mu^{(Z)}$ and diffusion $a^{(Z)} = \sigma^{(Z)} \sigma^{(Z)*}$ coefficients are affine functions of $z \in D$.*

So the general form of the dynamics of the factors of an exponential affine model is necessarily

$$dZ_t = (aZ_t + b)dt + \Sigma \begin{bmatrix} \sqrt{a_1 + b_1 \cdot Z_t} & 0 & \cdots & 0 \\ 0 & \sqrt{a_2 + b_2 \cdot Z_t} & \cdots & 0 \\ \vdots & \vdots & \vdots & \vdots \\ 0 & \cdots & 0 & \sqrt{a_k + b_k \cdot Z_t} \end{bmatrix} dW_t,$$

with initial condition $Z_0 \in D$. Here, a and Σ are $k \times k$ deterministic matrices, b, b_1, ..., b_k are k-dimensional deterministic vectors, and a_1, ..., a_k are scalars. The analysis of such a stochastic differential equation is very simple when all the b_i's are zero. In this case, there is existence and uniqueness of a solution, and the latter is a Gauss–Markov process. We shall see several examples later in this chapter. However, the situation is much more delicate when some of the b_i's are nonzero. Existence and uniqueness of solutions of such stochastic differential equations is not guaranteed, and analyzing the properties of this non-Gaussian diffusion is not easy because of the random volatility created by the nonzero b_i's. Again, we shall give the details for a couple of examples in what follows.

We were very casual in the way we ignored important technical assumptions, and in the informal way we stated the results on exponential affine models. The reader interested in proofs and complete statements of these important results is referred to the Notes & Complements at the end of the chapter.

Example 2.1. We specialize our discussion of the exponential affine models to a special case used most frequently. Let $Z_t^{(1)}, \ldots, Z_t^{(k)}$ be economic factors such that

$$r_t = \sum_{j=1}^{k} Z_t^{(j)}.$$

We suppose that these factors are independent so that

$$\begin{aligned} P(t,T) &= \mathbb{E}\left\{ \exp\left(-\int_t^T r_s ds \right) \Big| \mathcal{F}_t \right\} \\ &= \prod_{j=1}^{k} \mathbb{E}\left\{ \exp\left(-\int_t^T Z_s^{(j)} ds \right) \Big| \mathcal{F}_t \right\}. \end{aligned}$$

Notice that in what follows, we shall mostly concentrate on one-factor models for which $k = 1$ and $Z_t^{(1)} = r_t$ is the short rate. For the time being, we suppose that the factors are solutions of SDE

$$dZ_t^{(j)} = \mu_j(Z_t^{(j)})dt + \sigma_j(Z_t^{(j)})dw_t^{(j)}.$$

for independent Wiener processes $w^{(1)}, \ldots, w^k$. Now let us choose function coefficients μ_j and σ_j in such a way that they generate an affine term structure, say by letting

$$\mu_j(z) = \mu_j + \overline{\mu}_j z$$

and

$$\sigma_j(z) = \sqrt{\sigma_j + \overline{\sigma}_j z}.$$

Then we have

$$\mathbb{E}\left\{ \exp\left(-\int_t^T Z_s^{(j)} ds \right) \Big| \mathcal{F}_t \right\} = \exp\left(A^{(j)}(T-t) + B^{(j)}(T-t) Z_t^{(j)} \right)$$

for specific functions $A^{(j)}$ and $B^{(j)}$, which can be computed explicitly by solving the appropriate Riccati equation. Multiplying these equations together and taking the logarithmic derivative gives the formula for the forward rates:

$$f(t,T) = a(T-t) + b(T-t)Z_t$$

where

$$a(x) = -\sum_{j=1}^{k} \frac{d}{dx} A^{(j)}(x), \qquad \text{and} \qquad b(x) = -\left(\frac{d}{dx}B^{(1)}(x), \ldots, \frac{d}{dx}B^{(k)}(x)\right).$$

We can now solve for the factors by choosing k benchmark times to maturity $x_1, \ldots, x_k$ such that the matrix $\Gamma = [b^{(i)}(x_j)]_{i,j=1,\ldots,k}$ is invertible:

$$Z_t = \Gamma^{-1}(f(t,t+x_i) - a(x_i))_{i=1,\ldots,d}.$$

Hence, the economic factors in this model can be interpreted as a affine function of k benchmark forward rates. Furthermore, given the value of these k rates, all of the other rates $f(t,T)$ can be computed by the interpolation formula

$$f(t,T) = a(T-t) + b(T-t) \cdot \Gamma^{-1}(f(t,t+x.) - a(x.)).$$

2.3 Short Rate Models as One-Factor Models

This section is devoted to the particular case of the one-factor models, when the single factor is chosen to be the short interest rate. After all, if we have to limit ourselves to one factor, the short rate looks like a good choice. In this way, we recast the early models of the term structure based solely on the dynamics of the short rate, in the framework of factor models introduced earlier.

To conform with the standard notation, we write r_t for the single factor Z_t, and without any loss of generality we assume that the Wiener process W is scalar. In other words, we assume that the short interest rate r_t is the solution of a stochastic differential equation (SDE for short) of the form:

$$dr_t = \mu^{(r)}(t, r_t)\, dt + \sigma^{(r)}(t, r_t)\, dW_t \tag{2.13}$$

where the drift and volatility terms are given by real-valued (deterministic) functions

$$(t,r) \hookrightarrow \mu^{(r)}(t,r) \quad \text{and} \quad (t,r) \hookrightarrow \sigma^{(r)}(t,r)$$

such that existence and uniqueness of a strong solution hold. In most cases this will be guaranteed by assuming that these functions are uniformly Lipschitz. But as we shall see later, this sufficient condition is not satisfied by some of the most popular models. Recall the discussion of the affine models and their diffusion terms involving the square-root function (which is obviously not uniformly Lipschitz). We shall nevertheless review the main properties of the short interest models, both because of the important historical role they played, and because of the role they still play in many implementations of Monte Carlo pricing algorithms for risky, callable and/or convertible bonds. In any case, the existence and the uniqueness of a solution of the SDE (2.13) implies the Markov property of the short interest rate. As we shall see below, the Markov property of the factor vector (i.e. the state of the economy) is crucial for the derivation of the PDE used to compute derivative prices. Unfortunately, as the recent work of Ait-Sahalia [2] shows, the Markov property is not an assumption clearly supported by empirical studies. Notice that this Markov property assumption will not be part of our general models introduced later in Chap. 6. But in order to review the classical stochastic models of the short rate, we shall nevertheless work with it in this section.

2.3.1 Incompleteness and Pricing

Given such (Markovian) dynamics for the short interest rate, the money-market account $\{B_t;\, t \geq 0\}$ is defined as usual as the pay-off resulting for continuously compounding interest on a unit deposit:

$$\begin{cases} dB_t = r_t B_t \, dt, \\ B_0 = 1. \end{cases} \tag{2.14}$$

Equation (2.14) is a random ordinary differential equation (ODE for short) because the coefficient r_t is random. But it should not be viewed as an SDE because there is no Wiener process driving its randomness. For this reason, the solution:

$$B_t = \exp\left(\int_0^t r_s \, ds\right) \tag{2.15}$$

is still called the *risk-free asset* by analogy with the situations in which r_t is deterministic.

The next step is to price (zero coupon riskless) bonds or more general contingent claims by treating them as derivatives as in the Black–Scholes theory, the interest rate r_t playing the role of the underlying risky asset. But the analogy stops here. Indeed, the only tradable asset in such a market model is the money-market account (i.e. the risk free asset B_t) and it is not possible to form portfolios which can replicate interesting contingent claims, not even the zero coupon bonds. This means that such a market is not complete. The incompleteness of the model can be equivalently established by the non-uniqueness of the equivalent martingale measure. This fact is quite clear because, since B_t is the only tradable in our model, and since its discounted value $\tilde{B}_t = B_t^{-1} B_t \equiv 1$ is constant, it is always a martingale, hence

any equivalent measure $\mathbb{Q} \sim \mathbb{P}$ is an equivalent martingale measure !!

Because we assume that the filtration is generated by the Wiener process $\{W_t\}_{t \geq 0}$, any equivalent probability measure $\mathbb{Q}$ is determined by an Itô integrand $\{K_t;\, t \geq 0\}$, namely the adapted process whose Doleans exponential gives the Radon–Nykodym density of $\mathbb{Q}$ with respect to $\mathbb{P}$. In order to use Girsanov's theorem, we set:

$$\tilde{W}_t = W_t + \int_0^t K_s\, ds, \qquad \tilde{\mu}_t = \mu(t, r_t) - \sigma(t, r_t)K_t, \qquad \text{and} \qquad \tilde{\sigma}_t = \sigma(t, r_t) \tag{2.16}$$

because the process $\{\tilde{W}_t\}_t$ is a Wiener process for the probability structure given by $\mathbb{Q}$. On this new probability space the short rate r_t appears as the solution of the stochastic differential:

$$dr_t = \tilde{\mu}_t\, dt + \tilde{\sigma}_t\, d\tilde{W}_t. \tag{2.17}$$

Since the dynamics of r_t under $\mathbb{P}$ are not enough to price the bonds $P(t, T)$, even if we impose the no-arbitrage condition, pricing models based on the short interest rate r_t will have:

- either to specify the risk-premium process $\{K_t\}_t$ together with the (stochastic) dynamics of r_t under $\mathbb{P}$ as given by the drift and volatility processes μ_t and σ_t
- or to specify the (stochastic) dynamics of r_t under the risk-neutral probability measure $\mathbb{Q}$ by giving the risk adjusted drift and volatility processes $\tilde{\mu}_t$ and $\tilde{\sigma}_t$

Of these two equivalent prescriptions, most pricing models follow the latter. Because of that, the following remark is in order.

Remark 2.1. Statistical Estimation versus Calibration
If the model is given under the *historical* (also called *objective*) probability structure given by $\mathbb{P}$, historical data can or should be used to estimate the coefficients μ_t and σ_t. But historical data should not be used to estimate the coefficients $\tilde{\mu}_t$ and $\tilde{\sigma}_t$ when the model is specified under an equivalent martingale measure $\mathbb{Q}$. When pricing, the risk adjusted coefficients $\tilde{\mu}_t$ and $\tilde{\sigma}_t$ can only be inferred from existing prices. Indeed, in order to guess which equivalent martingale measure the market chose to get the prices for which we have quotes, we try to reverse engineer the process and get estimates of $\tilde{\mu}_t$ and $\tilde{\sigma}_t$ from these quotes. This is typically an ill-posed inverse problem whose solution requires the choice of a regularization procedure and an optimization algorithm. To distinguish it from the statistical estimation from historical data mentioned earlier, the term *calibration* is commonly used in practice.

2.3.2 Specific Models

All the models discussed below can be recast in one single equation for the risk-neutral dynamics (2.17). It reads:

$$dr_t = (\alpha_t - \beta_t r_t)\, dt + \sigma_t r_t^{\gamma}\, dW_t \tag{2.18}$$

where $t\hookrightarrow\alpha_t$, $t\hookrightarrow\beta_t$, and $t\hookrightarrow\sigma_t$ are deterministic non-negative functions of t and γ is a positive constant. These models have been studied in various degrees of generality and without any respect for the order in which they appeared historically, we list the special cases of interest. Note that these models are affine in the sense of the previous section whenever $\gamma = 0$ or $\gamma = 1/2$. Also for us, a case is of interest if explicit formulae and/or efficient computational procedures can be derived. The main features of these models include:

- When $\gamma \neq 0$, the volatility gets smaller as r_t approaches 0, allowing the drift term to be dominated by α_t and possibly preventing r_t from becoming negative.
- When $\beta_t > 0$, the drift term is a restoring force (similar to the Hooke's term appearing in mechanics) which always points toward the current mean value of α_t/β_t.
- The standard Lipschitz assumption of the strong (i.e. pathwise) existence and uniqueness result is not satisfied when $0 < \gamma < 1$. Nevertheless, existence and uniqueness still holds as long as $\gamma \geq 1/2$.

Bibliographic references are given in the Notes & Complements at the end of the chapter.

Vasicek Model

This model corresponds to the choices $\alpha_t \equiv \alpha$, $\beta_t \equiv \beta$ and $\sigma_t \equiv \sigma$ constant while $\gamma = 0$. So in this model, the risk adjusted dynamics of the short term rate are given by the SDE:

$$dr_t = (\alpha - \beta r_t)\, dt + \sigma\, dW_t. \tag{2.19}$$

The solution of this diffusion equation is a particular case of the processes of Ornstein–Uhlenbeck type discussed later in Sect. 4.5 of Chap. 4. Indeed, we can solve for the short rate explicitly

$$r_t = e^{-\beta t} r_0 + (1 - e^{-\beta t})\frac{\alpha}{\beta} + \int_0^t e^{-\beta(t-s)}\sigma dw_s.$$

It is clear from the above formula that such a process is Gaussian, and at each time $t > 0$ there is a positive probability that r_t is negative. This has been regarded by many as a reason not to use the Vasicek model. Despite this criticism, the model remained popular because of its tractability and because a judicious choice of the parameters can make this probability of negative interest rate quite small. Some even go as far as arguing that after all, real interest rates (i.e. rates adjusted for inflation) can be negative, and for this reason, the Vasicek model (2.19) is often used to model real interest rates.

The Cox–Ingersoll–Ross model

This model is often called the CIR model for short. It was introduced as an equilibrium model, but its claim of faith is attached to the desirable correction it brings to the Vasicek model: while keeping with the important mean reversion feature, by introducing a rate-level-dependent volatility, the resulting short rate will never become negative. This model corresponds to the choices $\alpha_t \equiv \alpha$, $\beta_t \equiv \beta$ and $\sigma_t \equiv \sigma$ constant as before in the Vasicek model, while now $\gamma = 1/2$. So, the risk adjusted dynamics of the short term rate are given by the SDE:

$$dr_t = (\alpha - \beta r_t)\,dt + \sigma\sqrt{r_t}\,dW_t. \tag{2.20}$$

The solution of this diffusion equation is sometimes called the square-root diffusion process. It was first studied by W. Feller who identified its transition probability as a non-central χ^2 distribution. It is not a Gaussian process, and for this reason, explicit formulae are usually more difficult to come by. See our discussion below. However, many positive features are preserved. It is still possible to use exact simulation like in the Gaussian case, and since the model is part of the family of affine models, explicit derivative pricing formulae are available.

The main shortcoming of the two short rate models presented above is with calibration to market data. We discuss this problem in Sect. 2.3.5 below.

Other Frequently Used Models

For the sake of completeness we quote some of the models introduced to overcome the shortcomings of the two most popular models introduced above.

- The **Dothan model** corresponds to the case $\alpha_t \equiv 0$, $\beta_t \equiv -\beta$ and $\sigma_t \equiv \sigma$ constant while $\gamma = 1$. In other words, the dynamics of the short term rate are given by the SDE:

  $$dr_t = \beta r_t\,dt + \sigma r_t\,dW_t. \tag{2.21}$$

  The solution of this diffusion equation is the classical geometric Brownian motion. It is given by the formula:

  $$r_t = r_0 e^{(\beta - \sigma^2/2)t + \sigma W_t}$$

  which shows that the short rate is always positive if it starts from $r_0 > 0$. The random variable r_t has a log-normal density, and many explicit formula can be derived from that fact.
  Unfortunately, the calibration problem suffers from the same shortcomings as the Vasicek and the CIR models.

- The **Black–Derman–Toy model** (BDT model for short) corresponds to the case $\alpha_t \equiv 0$ and $\gamma = 1$. The dynamics of the short term rate are given by the SDE:
$$dr_t = -\beta_t r_t \, dt + \sigma_t r_t \, dW_t. \tag{2.22}$$
An explicit formula for the solution of the above equation can be written as
$$r_t = r_0 e^{-\int_0^t (\beta_s + \sigma_s^2/2) ds + \int_0^t \sigma_s dW_s}.$$
In this model, the distribution of r_t is log-normal, so it is still almost surely non-negative. When compared to the Dothan model above, the loss due to the increased complexity of the distribution is barely compensated by the changes in the calibration issue. Indeed, the calibration problem remains over-determined when the time dependent volatility is chosen in a parametric family, but it becomes under-determined if it is estimated by nonparametric methods.
- The **Ho–Lee model** corresponds to the case $\beta_t \equiv 0$ and $\sigma_t \equiv \sigma$ constant while $\gamma = 0$. In other words, the dynamics of the short term rate are given by the SDE:
$$dr_t = \alpha_t \, dt + \sigma \, dW_t. \tag{2.23}$$
Again an explicit formula is available:
$$r_t = \int_0^t \alpha_s ds + \sigma W_t.$$
The comments made on the Black–Derman–Toy model apply as well to the current Ho–Lee model. Furthermore, the distribution of the random variable r_t is Gaussian, so there is a positive probability of that the interest rate becomes negative.

The last two models which we review can be seen as the culmination in the introduction of time dependent parameters in the Vasicek and CIR models. The need for time dependent models will be stressed once more in our discussion of the stiffness of the yield curves produced by these models, and of calibration issues.

- The **Vasicek–Hull–White model** corresponds to the case $\gamma = 0$. The dynamics of the short term rate are given by the SDE:
$$dr_t = (\alpha_t - \beta_t r_t) \, dt + \sigma_t \, dW_t. \tag{2.24}$$
As in the standard Vasicek model, the interest rate equation has an explicit solution:
$$r_t = e^{-\int_0^t \beta_s ds} r_0 + \int_0^t e^{-\int_s^t \beta_s ds} \alpha_s ds + \int_0^t e^{-\int_s^t \beta_s ds} \sigma_s dW_s.$$
Note that random variable r_t is Gaussian in the VHW model.

- Finally, the **CIR–Hull–White model** corresponds to the case $\gamma = 1/2$ in which case the dynamics of the short term rate are given by the SDE:

$$dr_t = (\alpha_t - \beta_t r_t)\, dt + \sigma_t \sqrt{r_t}\, dW_t. \tag{2.25}$$

 Just as with the standard CIR model, an explicit solution for r_t is not available; nevertheless, explicit bond pricing formulae can be found in some cases.

2.3.3 A PDE for Numerical Purposes

The analysis of the classical Black–Scholes–Merton derivative pricing theory has taught us that, since prices are given by expectations with respect to an equivalent martingale measure, they are solutions of a partial differential equation (PDE for short) whenever the underlying dynamics are given by a Markov process under the risk-neutral martingale measure. When the dimension of the factors underlying the derivatives is small, it makes sense to compute the prices by solving a PDE instead of computing expectations. This approach is still very popular and the purpose of this section is to review the derivation of these pricing PDE.

As before, we work with a fixed equivalent martingale measure $\mathbb{Q}$, and we assume that in the probability structure determined by this measure, the short rate is the unique solution to a stochastic differential (SDE for short) of the form:

$$dr_t = \mu(t, r_t)\, dt + \sigma(t, r_t)\, dW_t. \tag{2.26}$$

Notice that this assumption is slightly more restrictive than our original assumption (2.13) for the dynamics of the short rate under the objective probability measure $\mathbb{P}$. Indeed, on top of the difference in notation due to our dropping the tildes over the coefficients, the general stochastic differential under an equivalent measure given in (2.17) is not necessarily Markovian when the historical dynamics given under $\mathbb{P}$ are Markovian. This is so because the random process $\{\tilde{\mu}_t;\, t \geq 0\}$ may be a function of the whole past up to time t instead of being a deterministic function of r_t as in the form given by Eq. (2.26). Indeed, the risk adjustment process $\{K_t\}_t$ can be a function of the whole past.

Now, according to the no-arbitrage pricing paradigm, in such a model the price at time t of any contingent claim ξ with maturity T is given by the conditional expectation:

$$V_t = \mathbb{E}^{\mathbb{Q}}\{\xi e^{-\int_t^T r_s ds} | \mathcal{F}_t\}. \tag{2.27}$$

In particular, the price at time t of a zero coupon bond with maturity T is given by the formula:

$$P(t, T) = \mathbb{E}^{\mathbb{Q}}\{e^{-\int_t^T r_s ds} | \mathcal{F}_t\} \tag{2.28}$$

since in this case $\xi \equiv 1$. What we are about to say applies as well to any T-contingent claim whose pay-off is of the form $\xi = f(r_T)$. As a conditional expectation with respect to the past information available at time t, the price V_t, should be a function of the whole past of the Wiener process, or equivalently (under some mild assumption on the function $\sigma(t,r)$ which we shall not spell out) of the past information contained in the values of r_s for $0 \le s \le t$. But because of the choice of the model (2.26), the stochastic process r_t is Markovian, and the expectation of a random variable in the future at time t conditioned by the past given by r_s for $0 \le s \le t$ is as well the conditional expectation with respect to the present value r_t. In other words, formula (2.27) can be rewritten as:

$$\begin{aligned} V_t &= \mathbb{E}^{\mathbb{Q}}\{f(r_T)e^{-\int_t^T r_s ds}|r_s,\, 0 \le s \le t\} \\ &= \mathbb{E}^{\mathbb{Q}}\{f(r_T)e^{-\int_t^T r_s ds}|r_t\} \end{aligned} \tag{2.29}$$

which shows that V_t is in fact a deterministic function of t and r_t. Indeed, if we set:

$$F(t,r) = \mathbb{E}^{\mathbb{Q}}\{f(r_T)e^{-\int_t^T r_s ds}|r_t = r\} \tag{2.30}$$

then we have $V_t = F(t, r_t)$. We now explain why this function is the solution of a specific PDE. As in the Black–Scholes–Merton theory, we can use Itô's calculus and an arbitrage argument to derive this PDE. But since the arbitrage argument is captured by the use of the risk-neutral measure in pricing by expectation, this PDE can also be derived with the classical argument due to Feynman and Kac. Indeed, if we apply the Feynman–Kac formula to the Markov process $\{r_t;\, t \ge 0\}$ whose dynamics under $\mathbb{Q}$ are given by the stochastic differential equation (2.26), we get the following:

Proposition 2.2. *The no-arbitrage price at time t of any contingent claim ξ of the form $\xi = f(r_T)$ with maturity $T > t$ is of the form $F(t, r_t)$ where F is a solution of the parabolic equation:*

$$\frac{\partial F}{\partial t}(t,r) + \mu(t,r)\frac{\partial F}{\partial r}(t,r) + \frac{1}{2}\sigma(t,r)^2\frac{\partial^2 F}{\partial r}(t,r) - rF(t,r) = 0 \tag{2.31}$$

with the terminal condition $F(T,r) \equiv f(r)$.

Under mild regularity assumptions on the coefficients μ and σ (for example a global Lipschitz condition would do) this function $F(t,r)$ is the unique solution of the PDE (2.31) which satisfies the terminal condition $F(T,r) \equiv f(r)$. In particular, the price at time t when the spot rate is equal to r of a zero coupon bond with maturity T is given by the solution of the PDE (2.31) with the terminal condition $F(T,r) \equiv 1$ since the pay-off is $\xi \equiv 1$. Pricing bonds and interest rate derivatives by solving numerically such a PDE is not uncommon.

The PDE (2.31) can easily be solved numerically, either by an explicit finite-difference scheme, or by by a standard implicit scheme if one worries about stability issues.

2.3.4 Explicit Pricing Formulae

We now reap the benefits from the special features of the affine models introduced at the start of the chapter. We give explicit formulae for the prices of the zero coupon bonds computed in the Vasicek and CIR models.

Vasicek Model

The pricing PDE reads:

$$\frac{\partial P}{\partial t} + \frac{1}{2}\sigma^2 \frac{\partial^2 P}{\partial r^2} + (\alpha - \beta r)\frac{\partial P}{\partial r} - rP = 0 \tag{2.32}$$

with the constant function 1 as terminal condition. So for a zero coupon bond the solution is of the form:

$$P_V(t, T, r) = e^{A(T-t)+B(T-t)r}$$

with

$$B(x) = -\frac{1}{\beta}\left(1 - e^{-\beta x}\right)$$

and

$$A(x) = \frac{4\alpha\beta - 3\sigma^2}{4\beta^3} + \frac{\sigma^2 - 2\alpha\beta}{2\beta^2}x + \frac{\sigma^2 - \alpha\beta}{\beta^3}e^{-\beta x} - \frac{\sigma^2}{4\beta^3}e^{-2\beta x}.$$

These facts can be proven by plugging the expressions for $A(x)$ and $B(x)$ in $P_V(t, T, r)$, and checking that the latter satisfies the partial differential equation (2.32) with the right terminal condition. However, these formulae can be derived directly by computing the conditional expectation (2.29) using the expression of the Laplace transform of a Gaussian random variable.

It is sometimes easier to deal with the forward rates $f(t, T) = -\frac{\partial}{\partial T}\log P(t, T)$, rather than the bond prices directly. Define the function f_V by the formula

$$f_V(t, T, r) = re^{-\beta(T-t)} + \frac{\alpha}{\beta}\left(1 - e^{-\beta(T-t)}\right) - \frac{\sigma^2}{2\beta^2}\left(1 - e^{-\beta(T-t)}\right)^2. \tag{2.33}$$

Then the forward rates for the Vasicek model are given by $f(t, T) = f_V(r, T, r_t)$.

CIR Model

The pricing PDE reads:

$$\frac{\partial P}{\partial t} + \frac{1}{2}\sigma^2 r \frac{\partial^2 P}{\partial r^2} + (\alpha - \beta r)\frac{\partial P}{\partial r} - rP = 0 \tag{2.34}$$

with the constant function 1 as terminal condition. So for a zero coupon bond the price is also of the form:

$$P_{CIR}(t,T,r) = e^{A(T-t)+B(T-t)r}$$

with

$$B(x) = -\frac{2(e^{\gamma x}-1)}{(\gamma+\beta)e^{\gamma x}+(\gamma-\beta)}$$

and

$$A(x) = \frac{\alpha\beta(\gamma+\beta)}{\sigma^2}x - \frac{\alpha\beta}{2\sigma^2}\log\left(\frac{(\gamma+\beta)e^{\gamma x}+(\gamma-\beta)}{2\gamma}\right)$$

with

$$\gamma = \sqrt{\beta^2+2\sigma^2}.$$

As before, these facts can be proven by plugging the expressions for $A(x)$ and $B(x)$ in $P_{CIR}(r,t,T)$, and checking that the latter satisfies the partial differential equation (2.34) with the right terminal condition. They can also be derived by solving the Riccati equation.

The forward rates for the CIR model are given by

$$f_{CIR}(t,T,r) = \frac{4\gamma^2 e^{\gamma(T-t)}}{[(\gamma+\beta)e^{\gamma(T-t)}+(\gamma-\beta)]^2}r + \frac{2\alpha(e^{\gamma(T-t)}-1)}{(\gamma+\beta)e^{\gamma(T-t)}+(\gamma-\beta)}. \tag{2.35}$$

VHW Model

The bond prices and forward rates can be computed explicitly for the VHW model.

$$f_{VHW}(t,T,r) = re^{-\int_t^T \beta_s ds} + \int_t^T \alpha_s e^{-\int_s^T \beta_u du} ds$$
$$- \int_t^T \int_t^y \sigma_s^2 e^{-\int_s^T \beta_u du - \int_s^y \beta_u du} ds\, dy$$

The above formula can be derived from the fact that the interest rate in the VHW model is a Gaussian process. The special case when the mean-reverting parameter $\beta_t \equiv \beta$ and the volatility $\sigma_t \equiv \sigma$ are constant is particularly interesting:

$$f_{VHW}(t,T,r) = re^{-\beta(T-t)} + \int_t^T \alpha_s e^{-\beta(T-s)} ds - \frac{\sigma^2}{2\beta^2}\left(1-e^{-\beta(T-t)}\right)^2. \tag{2.36}$$

CIRHW Model

The bond prices and forward rates can be computed explicitly for the CIRHW model, at least in the special case when the mean-reverting parameter $\beta_t \equiv \beta$ and the volatility $\sigma_t \equiv \sigma$ are contant. The formula is

$$f_{CIRHW}(t,T,r) = \frac{4\gamma^2 e^{\gamma(T-t)}}{[(\gamma+\beta)e^{\gamma(T-t)} + (\gamma-\beta)]^2} r + \int_t^T \frac{4\gamma^2 e^{\gamma(T-s)}\alpha_s}{[(\gamma+\beta)e^{\gamma(T-s)} + (\gamma-\beta)]^2} ds.$$

2.3.5 Rigid Term Structures for Calibration

The Vasicek and the CIR models depend upon the three parameters α, β, and σ. Three quoted prices are often enough to determine these parameters. But many more prices are available and it is not clear how to choose three prices out of the bunch, especially because most likely, any given set of three prices will give a different set of values to the parameters α, β and σ.

The calibration problem is very often over-determined for parametric models with a small number of parameters. An approximate solution needs to be chosen: least squares is usually considered as a reasonable way out. Notice that we are in the same situation as in Chap. 1 when trying to estimate the term structure with a parametric family of curves such as the Nelson–Siegel or the Svennson families.

The above formulae for the zero coupon bonds in the Vasicek and CIR models can be used to compute and plot the term structure of interest rate. Tweaking the parameters can produce yield curves with one hump or one dip, but it is very difficult (if not impossible) to calibrate the parameters so that the hump/dip sits where desired. There are not enough parameters to calibrate the models to account for observed features contained in the prices quoted on the markets. Recall Fig. 1.2 of Chap. 1 for example.

This undesirable rigidity of the yield curves attached to the short rate models leads to the introduction of the time dependent models (also called evolutionary models) reviewed in Sect. 2.3.2 above. For instance, by choosing the drift $t \hookrightarrow \alpha_t$ appropriately, the VHW and CIRHW models can be made to match any initial forward curve $T \hookrightarrow f(0,T)$. In this way, the model is made compatible with the current *observed* forward curve. However, the next time we check the forward curve given by the market, it will presumably not agree with the forward curve implied by the model, hence the need to recalibrate *each time*. This constant need for recalibration is a good reason to lose faith (if not trust) in the model as the latter appears as a *one-period* model. This is at odds with the original belief that we had a dynamical model capable of being used over time!

Another explanation for the difficulties of enabling the short rate models to be consistent with the daily changes of the forward curve goes as follows.

From the point of view of the forward curves, specifying a short rate model amounts to specifying the (stochastic) dynamics of the whole forward curve by specifying the (stochastic) dynamics of the left-hand point of the curve (remember that $r_t = f(t,t)$). The motion of a curve should have a continuum of degrees of freedom, and being able to specify only one of them to determine this motion should lead to another kind of rigidity.

Indeed, pick two maturities T_1 and T_2 and consider the two-dimensional random vector $(f(t,T_1), f(t,T_2))$. In a short rate model for the term structure, the support of this random vector is contained in the closure of the set

$$S = \{(f^{(r)}(t,T_1,r), f^{(r)}(t,T_2,r)) : r \in \mathcal{D}\}$$

where $\mathcal{D}$ is typically $\mathbb{R}$ or the halfline $\mathbb{R}_+$. In most cases, the set $\bar{S}$ is a Lebesgue measure zero subset of $\mathbb{R}^2$, in which case the law of $(f(t,T_1), f(t,T_2))$ does not have a density.

Furthermore, in a short rate model, the forward rates $f(t,T_1)$ and $f(t,T_2)$ move in lock step. This fact can be illustrated by computing the statistical correlation between the increments of the forward rates. It is easy to see that the correlation coefficient between the "random variables" $df(t,T_1)$ and $df(t,T_2)$ is necessarily equal to 1!

2.4 Term Structure Dynamics

Starting from a factor model, we noticed that the bond prices were necessarily Itô processes with stochastic dynamics of the form (2.5). There is no arbitrage in the model if there is an equivalent martingale measure $\mathbb{Q}$, such that all discounted bond prices $\{\tilde{P}(t,T)\}_{t\in[0,T]}$ are local martingales, where the discounted bond price at time t for maturity T is given by

$$\tilde{P}(t,T) = e^{-\int_0^t r_s ds} P(t,T).$$

Since we are working with a filtration generated by a d-dimensional Wiener process, any local martingale can be written as a stochastic integral with respect to this multivariate Wiener process and we have

$$d\tilde{P}(t,T) = \sum_{i=1}^{d} \tau^{(i)}(t,T) dw_t^{(i)} \tag{2.37}$$

for some predictable processes $\{\tau^{(i)}(t,T)\}_{t\in[0,T]}$.

2.4.1 The Heath–Jarrow–Morton Framework

We now introduce a framework which will play a central role in our analysis of term structure models. This framework will be studied in much detail in Chap. 6.

Generalizing the finite factor models as hinted above, we consider a model such that the discounted bond prices $\{\tilde{P}(t,T)\}_{t\in[0,T]}$ are continuous local martingales simultaneously for all T. Of course, such a market model is free of arbitrage opportunities. That is, we essentially take Eq. (2.37) as our starting point.

In fact, in the framework proposed by Heath, Jarrow, and Morton [80] the forward rates $\{f(t,T)\}_{t\in[0,T]}$ are assumed to be Itô processes for each T with dynamics given by

$$df(t,T) = \alpha(t,T)dt + \sum_{j=1}^{d} \sigma^{(j)}(t,T)dw_t^{(j)}, \tag{2.38}$$

where for each j and T the process $\{\sigma^{(j)}(t,T)\}_{t\in[0,T]}$ is assumed to be predictable with respect to the filtration generated by the Wiener process and where the drift is given by the formula

$$\alpha(t,T) = \sum_{j=1}^{d} \sigma^{(j)}(t,T) \int_t^T \sigma^{(j)}(t,s)ds.$$

The above formula for the drift in terms of the volatilities was discovered by Heath, Jarrow, and Morton [80] and is commonly called the HJM drift condition.

We shall refer to any term structure model which has the property that the forward rates simultaneously satisfy stochastic differential equations of the form of Eq. (2.38) as a *finite rank HJM model.* The adjective "finite rank" indicates that the Wiener process $\{W_t\}_{t\geq 0}$ is finite dimensional; in Chap. 6 we will consider abstract HJM models driven by a Wiener process taking values in an infinite dimensional space.

As we noted in the previous section, if the bond prices are modeled as a deterministic function of a finite dimensional diffusion $\{Z_t\}_{t\geq 0}$, then the stochastic dynamics of the discounted bond prices are necessarily of the form of Eq. (2.37). Therefore all of the finite factor models studied in this chapter are finite rank models. The converse is false in general.

To contract the HJM approach to the factor approach considered before, notice that the state variable is taken to be the entire forward rate curve $T\hookrightarrow f(t,T)$ rather than the finite dimensional vector Z_t of economic factors. In particular, whereas the factor models are determined by the following data: the functions $\mu^{(Z)}$ and $\sigma^{(Z)}$ and the initial condition Z_0, an HJM model is determined by the stochastic processes $\{\sigma(t,T)\}_{t\in[0,T]}$ and the initial forward curve $f(0,\cdot)$. At this level of generality, we have a lot of freedom in choosing the volatility processes $\{\sigma(t,T)\}_{t\in[0,T]}$. One way to choose the volatility is to assume that it is of the form $\sigma(t,T) = \sigma^{(f)}(t,T,f(t,\cdot))$. This approach is described for the class of abstract HJM models studied in Chap. 6. In what follows, we do not make such an assumption.

2.4.2 Hedging Contingent Claims

We now consider the problem of hedging an interest rate contingent claim in the context of a finite rank HJM model. We assume that the measure $\mathbb{Q}$ is the unique measure for which the discounted bond prices $\{\tilde{P}(t,T)\}_{t\in[0,T]}$ are local martingales simultaneously for all T. In particular, we consider models where the discounted bond prices have stochastic dynamics given by

$$d\tilde{P}(t,T) = \sum_{i=1}^{d} \tau^{(i)}(t,T)dw_t^{(i)}.$$

Note that the discounted bond prices are given by an infinite number of stochastic differential equations, one for each value of T, but they are all driven by the same finite dimensional Wiener process $\{W_t\}_{t\geq 0}$.

Besides the fact that the mathematics of finite dimensional Wiener processes is easier to handle than that of infinite dimensional ones, the assumption that $\{W_t\}_{t\geq 0}$ is finite dimensional can be justified by appealing to the statistics of the yield and forward rate curves observed on the market. Indeed, the principal component analysis cited in Chap. 1 lends credence to term structure models driven by a Wiener process of dimension three or four.

Consider the problem of replicating the real $\mathcal{F}_T$-measurable random variable ξ corresponding to the payout of an interest rate contingent claim that matures at a fixed time $T > 0$. We choose as our hedging instruments the set of zero coupon bonds and the risk free bank account process $\{B_t\}_{t\geq 0}$, where as always $B_t = e^{\int_0^t r_s\, ds}$.

Pick d dates $T_1 < T_2 < \cdots < T_d$ with $T_1 > T$ and note that the d-dimensional vector of discounted bond prices $(\tilde{P}(t,T_1),\ldots,\tilde{P}(t,T_d))$ has risk neutral dynamics given by the stochastic differential equation

$$d\tilde{P}(t,T_i) = \sum_{j=1}^{d} \tau^{(j)}(t,T_i)\; dw_t^{(j)}. \tag{2.39}$$

If the $d \times d$ matrix-valued random variable σ_t given by

$$\sigma_t = \left[\tau^{(j)}(t,T_i)ds\right]_{i,j=1,\ldots,d} \tag{2.40}$$

is invertible for almost all $(t,\omega) \in [0,T]\times\Omega$, the model given by Eq. (2.39) is of a complete market consisting of d risky assets. For this finite rank model, the theory of contingent claim replication is well-known; we will see that we need only apply the martingale representation theorem to the discounted payout $B_T^{-1}\xi = \tilde{\xi}$ to compute the hedging strategy.

Consider a strategy such that at time t the portfolio consists of ϕ_t^i units of the bond with maturity T_i for $i = 1,\ldots,d$ and of ψ_t units of the bank account.

As usual, we insist that our wealth process $\{X_t = \langle\phi_t, P_t\rangle + \psi_t B_t\}_{t\geq 0}$ satisfies the self-financing condition

$$dX_t = \langle\phi_t, dP_t\rangle + \psi_t dB_t$$

where $\phi_t = (\phi_t^1, \ldots, \phi_t^d)$ is the vector of portfolio weights and $P_t = (P_t(T_1), \ldots, P_t(T_k))$ is the vector of bond prices. We now show that there exist processes $\{\phi_t\}_{t\in[0,T]}$ and $\{\psi_t\}_{t\in[0,T]}$ such that $X_T = \xi$ almost surely.

By Eqs. (2.39) and (2.40), the dynamics of the vector of discounted bond prices are given by $d\tilde{P}_t = \sigma_t dW_t$, and consequently, the dynamics of the discounted wealth process are given by

$$d\tilde{X}_t = \langle\phi_t, d\tilde{P}_t\rangle = \langle\sigma_t^*\phi_t, dW_t\rangle.$$

On the other hand, if $\mathbb{E}\{\tilde{\xi}^2\} < +\infty$, we can apply Itô's martingale representation theorem to conclude that there exists a d-dimensional adapted process $\{\alpha_t\}_{t\in[0,T]}$ such that $\mathbb{E}\left\{\int_0^T \|\alpha_t\|^2 dt\right\} < +\infty$ and

$$\tilde{\xi} = \mathbb{E}\{\tilde{\xi}\} + \int_0^T \langle\alpha_t, dW_t\rangle.$$

Setting the initial wealth $X_0 = \mathbb{E}\{\tilde{\xi}\}$ and portfolio weights $\phi_t = \sigma_t^{*-1}\alpha_t$ and $\psi_t = \tilde{X}_t - \langle\phi_t, \tilde{P}_t\rangle$ we find our desired replicating strategy.

We see then that for every claim ξ satisfying an appropriate integrability condition, there exists a self-financing portfolio consisting of bonds with maturities $T_1, T_2, \ldots, T_d$ and the bank account replicating the payoff of the contingent claim ξ. This is quite in line with the intuition developed from the Black–Scholes theory which taught us that in order to hedge all reasonable claims, we need only as many tradable assets as there are independent Wiener processes.

Notice that the above argument does not depend on a Markov assumption. For instance, the discounted bond prices $\{\tilde{P}_t = (\tilde{P}(t,T_1), \ldots, \tilde{P}(t,T_d)\}_{t\in[0,T]}$ need not be a Markov process. Loosely speaking, all that we have assumed is that the increment dW_t of the Wiener process can be recovered from knowledge of the increment $d\tilde{P}_t$ of the discounted bond prices.

2.4.3 A Shortcoming of the Finite-Rank Models

The assumption that the driving noise is finite-dimensional has an annoying implication: There typically exist hedging strategies which are rather unrealistic from the point of view of a fixed income trader.

The dates $T_1, \ldots, T_d$ in the above discussion were chosen *arbitrarily*; that is to say, the finite-rank assumption leads to the unrealistic situation that the hedging instruments can be chosen independently of the claim to be hedged.

For instance, consider the problem of hedging a call option on a bond of maturity five years in the context of an HJM model driven by three independent Wiener processes. According to the theory presented above, a portfolio of bonds of maturities 20, 25, and 30 years and the bank account could perfectly hedge the option. We are left with a puzzle: Why does our intuition suggest that a trader would prefer a hedging portfolio consisting of bonds with maturities closer to five years rather than with the above portfolio when the theory predicts that both strategies are just as good?

The shortcoming of finite-rank models is that, although there are bonds of very many maturities available to trade, most of these bonds are redundant. Indeed, the increment of the discounted bond price for a given maturity can be expressed as a linear combination of the increments of the discounted bond prices of d arbitrarily chosen maturities. It seems that a more intuitively satisfying model of the interest term structure would somehow incorporate a notion of *maturity-specific risk*. Such a model would have the following two desirable features:

- If a claim ξ can be hedged by a portfolio of zero coupon bonds, then the hedging strategy is unique.
- The maturities of the bonds used as hedging instruments for ξ depend on the maturities of the bonds underlying ξ.

In particular, a model which exhibits maturity-specific risk has the property that the increments of discounted bond prices of any finite set maturities are linearly independent. In Chap. 6 we will see that such models do in fact exist.

A natural first step to building a better model would be to recognize the infinite dimensional character of the term structure. This would entail rewriting the dynamics as an evolution equation in an infinite dimensional space, for instance a separable Hilbert space. In this new framework we would like to let the dimension of the driving Wiener process be infinite to provide the source of maturity-specific risk and to resolve the issue of non-uniqueness of hedging strategies discussed here. We would also like the resulting model to be consistent with the principal component analysis of the term structure. We address these issues in the last chapter devoted to the generalized models involving possibly infinitely many independent Wiener processes. But before we can take $d = \infty$ in our model equations, we need to develop analysis tools capable of handling infinitely many driving Wiener processes. This is the purpose of the following chapters.

2.4.4 The Musiela Notation

We now rewrite the equation for the dynamics of the instantaneous forward rates, viewing terms in the original equation as functions of the maturity date T, as restating the equality for all T's as an equality between functions.

We get:

$$f(t,\,\cdot\,) = f(0,\,\cdot\,) + \int_0^t \alpha(s,\,\cdot\,)ds + \sum_{j=1}^{d}\int_0^t \sigma^{(j)}(s,\,\cdot\,)dw_s^{(j)} \tag{2.41}$$

or in differential form:

$$df(t,\,\cdot\,) = \alpha(t,\,\cdot\,)dt + \sum_{j=1}^{d}\sigma^{(j)}(t,\,\cdot\,)dw_t^{(j)}. \tag{2.42}$$

This form is screaming for an interpretation as an equation for the dynamics of a function of the variable T. Unfortunately, for different t's, the functions $f(t,\,\cdot\,)$ are objects of different nature since they have different domains of definition.

As we already pointed out, the way out is to reparameterize the forward curve by the time to maturity $x = T - t$. In this way,

$$f_t(x) = f(t, t+x), \qquad\qquad t \geq 0,\ x \geq 0, \tag{2.43}$$

the forward curve at time t becomes a function $f_t : x \hookrightarrow f_t(x)$ with a domain independent of t, say the interval $[0, x_{\max}]$ with possibly $x_{\max} = \infty$. We shall propose in Chap. 6 several function spaces F to accommodate these functions of x, but in the mean time we may think of the space F as a subspace of space $C[0, x_{\max}]$ of continuous functions. Rewriting the integral form (2.41) of the model using the notation (2.43) we get:

$$\begin{aligned} f_t &= f(0, t+\,\cdot\,) + \int_0^t \alpha(s, t+\,\cdot\,)ds + \sum_{j=1}^{d}\int_0^t \sigma^{(j)}(s, t+\,\cdot\,)dw_s^{(j)} \\ &= S_t f_0 + \int_0^t S_{t-s}\alpha_s\,ds + \sum_{j=1}^{d}\int_0^t S_{t-s}\sigma_s^{(j)}\,dwi_s^{(j)} \end{aligned} \tag{2.44}$$

provided we set:

$$\alpha_t : x \hookrightarrow \alpha_t(x) = \alpha(t, t+x) \quad \text{and} \quad \sigma_t^{(j)} : x \hookrightarrow \sigma_t^{(j)}(x) = \sigma^{(j)}(t, t+x).$$

and provided the notation S_t is used for the left shift operator defined by

$$[S_t f](x) = f(x+t).$$

So the HJM prescription (2.41) for the dynamics of the forward curve appears as an integral evolution equation in infinite dimensions, given by a stochastic differential equation in a function space. Differentiating both sides of (2.41) with respect to t we get (at least formally):

$$df_t = \left(\frac{d}{dx}f_t + \alpha_t\right)dt + \sum_{j=1}^{d}\sigma_t^{(j)}\,dw_t^{(j)}.$$

The differential operator $A = d/dx$ complicates things somehow. Indeed, it may not be defined everywhere in F since F could contain non-differentiable functions. In other words, it is possibly unbounded as an operator on F. The reason for the seemingly sudden appearance of the differential operator A in the stochastic differential Eq. (2.42) should be clear: replacing T by $T = t+x$ forces us to take a derivative of f with respect to its second variable when we compute the differential with respect to t. For most of the natural choices of the space F, such an operator cannot be defined everywhere on F. In other words, A will presumably be an unbounded operator defined on a domain $\mathcal{D}(A)$ which will be at best a dense subspace of F.

Since the differential form (2.42) is always more singular than its integral analog, we will try to base the analysis of the model on the latter.

2.4.5 Random Field Formulation

Let us define the random field $\{Z_t(x);\, t \geq 0,\, x \in [0, x_{\max}]\}$ by:

$$Z_t(x) = \int_0^t \sum_{j=1}^d \frac{\sigma_s^{(j)}(x)}{\overline{\sigma}(x)} dw_s^{(j)}$$

with:

$$\overline{\sigma}_t(x) = \sqrt{\sum_{j=1}^d \sigma_s^{(j)}(x)^2}.$$

The random field $Z_t(x)$ is a very interesting object. Indeed, for each fixed time to maturity $x \in [0, x_{\max}]$, it is a Wiener process (it is a martingale and everything was done to make sure that its quadratic variation was t). In essence, for any given (fixed) time to maturity x, the values of the random field $Z_t(x)$ give the random kicks driving the (stochastic) dynamics of the instantaneous forward rate $f_t(x)$ with time to maturity x. Indeed, the dynamic equation can be rewritten in the form:

$$df_t(x) = \left(\frac{d}{dx} f_t(x) + \alpha_t(x)\right) dt + \overline{\sigma}_t(x) dZ_t(x), \tag{2.45}$$

and, except for the possible coupling due to the differential operator, the Eqs. (2.45) appear as a system of stochastic equations of Itô's type, one equation per time to maturity x, each equation being driven by the Wiener process $\{Z_t(x)\}_{t\geq 0}$. However, this statement is deceivingly simple, for the structure of $\{Z_t(x)\}_{t\geq 0}$ can be very complex: it captures most of the dependence between the forwards with different maturities, and in general, the random variables $Z_t(x_1)$ and $Z_t(x_2)$ are not jointly Gaussian if x_1 and x_2 are different. Nevertheless, several authors have taken Eq. (2.45) as the starting point for modeling the forward rates as a random field. One of the goals of the remainder of this book is to understand rigorously such random field models.

Notice that, in the case of a one-factor model, assuming that $\sigma^{(1)}(x) \geq 0$, then with $\overline{\sigma}_t(x) = \sigma^{(1)}(x)$ and $w_t = w_t^{(1)}$, Eq. (2.45) reads:

$$df_t(x) = \left(\frac{d}{dx} f_t(x) + \alpha_t(x) \right) dt + \overline{\sigma}_t(x) dw_t,$$

which shows clearly that the simultaneous motions of all the instantaneous forward rates are driven by the very same Wiener process $\{w_t\}_{t \geq 0}$. This is the source of stiffness already mentioned.

2.5 Appendices

Martingale Measures and Arbitrage

This first appendix is intended to provide a very brief introduction to the dynamic theory of asset prices. We assume that the financial market is given by a $d+1$-dimensional stochastic process $\{(B_t, P_t^{(1)}, \ldots, P_t^{(d)})\}_{t \geq 0}$. The components of the market process represent the time evolution of the prices of financial instruments. We distinguish the positive process $\{B_t\}_{t \geq 0}$ which represents the value of a bank account accumulating interest at the spot interest rate.

A trading strategy is a $d+1$-dimensional stochastic process $\{(\psi_t, \phi_t^{(1)}, \ldots, \phi_t^{(d)})\}_{t \geq 0}$. The wealth at time $t \geq 0$ of an investor employing such a strategy is given by the formula:

$$X_t = \psi_t B_t + \sum_{j=1}^{d} \phi_t^{(j)} P_t^{(j)} \tag{2.46}$$

where the random variable $\phi_t^{(j)}$ represents the number of shares of the j-th asset held by the investor, and the product $\psi_t B_t$ represents the portion of wealth held in the bank account.

In order to allow for trading in this market, we need to introduce a notion of available information. Indeed, market participants are not clairvoyant and can only make trading decisions based on information available today. The notion of information is formalized by the probabilistic concept of a filtration. Let $(\Omega, \mathcal{F}, \mathbb{P})$ be the probability space on which the market process is defined, and let $\{\mathcal{F}_t\}_{t \geq 0}$ be a filtration satisfying the usual assumptions and such that the market process is adapted.

In this book, we consider exclusively trading strategies which are *self-financing*. That is, the investor has no external income or expenses, and the changes in the wealth are due only to the fluctuations in the asset prices. First consider a simple predictable trading strategy $\{(\psi_t, \phi_t^{(1)}, \ldots, \phi_t^{(d)})\}_{t \geq 0}$

where each component has the representation

$$\phi_t^{(j)} = \sum_{i=1}^{n} \phi_{t_i}^{(j)} \mathbf{1}_{(t_j,t_{j+1}]}(t)$$

for a deterministic set of times $0 \leq t_1 < \ldots < t_n$ and where each $\phi_{t_i}^{(j)}$ is $\mathcal{F}_{t_i}$-measurable. The self-financing condition then becomes

$$X_{t_{i+1}} - X_{t_i} = \psi_{t_i}(B_{t_{i+1}} - B_{t_i}) + \langle \phi_t, P_{t_{i+1}} - P_{t_i} \rangle \tag{2.47}$$

where $\langle \cdot, \cdot \rangle$ denotes the standard Euclidean scalar product on $\mathbb{R}^d$. Solving for ψ_{t_i} in Eq. (2.46) and inserting the result in Eq. (2.47) yields

$$\tilde{X}_{t_{i+1}} - \tilde{X}_{t_i} = \langle \phi_t, \tilde{P}_{t_{i+1}} - \tilde{P}_{t_i} \rangle \tag{2.48}$$

where $\tilde{X}_t = B_t^{-1} X_t$ denotes the *discounted wealth* and $\tilde{P}_t = B_t^{-1} P_t$ denotes the *discounted asset prices*. The effect of the above algebraic manipulation is to change the numeraire from units of currency into units of the bank account.

Equation (2.48) shows that in the limit of continuous trading, the discounted wealth satisfies the Itô stochastic differential equation

$$d\tilde{X}_t = \langle \phi_t, d\tilde{P}_t \rangle$$

or more precisely, the wealth is given by the Itô stochastic integral

$$\tilde{X}_t = \tilde{X}_0 + \int_0^t \langle \phi_s, d\tilde{P}_s \rangle.$$

In order for the above stochastic integral to be well-defined, we assume that the discounted asset price process is a semi-martingale and that the trading strategy is a predictable process satisfying an appropriate integrability condition.

We need to impose another condition on the trading strategy in order to develop an economic meaningful theory. Indeed, since we are working in continuous time, there are pathological doubling strategies which promise arbitrarily large gains almost surely in finite time. The problem with such strategies is that the investor must have an infinite credit line, since he may go very deep into debt while employing the doubling strategy. To remedy the situation, we introduce the concept of an admissible strategy. There is more than one way to do this, but for the sake of this appendix, we offer this definition:

Definition 2.1. *A trading strategy* $\{\phi_t\}_{t \geq 0}$ *is* admissible *if the stochastic integral* $\int_0^t \langle \phi_s, d\tilde{P}_s \rangle$ *is bounded from below uniformly in* $t \geq 0$ *and* $\omega \in \Omega$.

We will in fact use a different definition of admissible strategy in our discussion of bond portfolios in Chap. 6. But in any case, now that the groundwork is laid, we can define a central concept to the theory.

Definition 2.2. *An* arbitrage *is admissible trading strategy* $\{\phi_t\}_{t\geq 0}$ *such that*

$$\mathbb{P}\left\{\int_0^T \langle \phi_s, d\tilde{P}_s\rangle \geq 0\right\} = 1$$

and

$$\mathbb{P}\left\{\int_0^T \langle \phi_s, d\tilde{P}_s\rangle > 0\right\} > 0$$

for some time $T > 0$.

What follows is a simple version of the so-called Fundamental Theorem of Asset Pricing.

Theorem 2.1. *There are no arbitrage strategies if there exists a probability measure* $\mathbb{Q}$*, equivalent to* $\mathbb{P}$*, such that the discounted asset prices process* $(\tilde{P}_t)_{t\geq 0}$ *is a local martingale for* $\mathbb{Q}$.

Recall that measures $\mathbb{P}$ and $\mathbb{Q}$ on the measurable space $(\Omega, \mathcal{F})$ are equivalent if they share the same null events. That is, if $\mathbb{P}$ and $\mathbb{Q}$ are equivalent then $\mathbb{P}(E) = 0$ if and only if $\mathbb{Q}(E) = 0$.

Proof. Suppose there exists an equivalent measure $\mathbb{Q}$ such that $(\tilde{P}_t)_{t\geq 0}$ is a local martingale. Let $\{\phi_t\}_{t\geq 0}$ be an admissible strategy such that

$$\int_0^T \langle \phi_s, d\tilde{P}_s\rangle \geq 0$$

$\mathbb{P}$-a.s. for some $T > 0$. Since $\mathbb{Q}$ is equivalent to $\mathbb{P}$, the above inequality holds $\mathbb{Q}$-a.s. also. Since $(\tilde{P}_t)_{t\geq 0}$ is a local martingale for $\mathbb{Q}$ and $\{\phi_t\}_{t\geq 0}$ is admissible, the process $\{\int_0^t \phi_s, d\tilde{P}_s\}_{t\geq 0}$ is a supermartingale for $\mathbb{Q}$. The following inequality holds:

$$\mathbb{E}^{\mathbb{Q}}\left\{\int_0^T \langle \phi_s, d\tilde{P}_s\rangle\right\} \leq 0$$

The stochastic integral is therefore zero $\mathbb{Q}$-a.s. and thus $\mathbb{P}$-a.s. In particular, the strategy $\{\phi_t\}_{t\geq 0}$ is not an arbitrage. □

The converse of this theorem is generally not true, strictly speaking, since the notion of arbitrage used here is too strong. There has been much work in identifying the "right" notion of arbitrage so that the lack of arbitrage is equivalent to the existence of a martingale measure. See the Notes & Complements for details.

Nevertheless, since the existence of an equivalent martingale measure is so closely related to the lack of arbitrage in the market, we shall often blur the distinction between these concepts. In particular, when we say that a market has no arbitrage, we mean the stronger statement that there exists an equivalent measure under which the discounted asset prices are all local martingales.

No Arbitrage in Factor Models

The absence of arbitrage in factor models can be made explicit in the form of a *drift condition* by applying Itô's formula to the function $P^{(Z)}(t,T,Z_t)$ giving the bond prices $P(t,T)$ in (2.1), and identifying the resulting drift to the short interest rate. For example, in the homogeneous case where $\mu^{(Z)}(t,z) = \mu^{(Z)}(z)$ and $\sigma^{(Z)}(t,z) = \sigma^{(Z)}(z)$, and the functions $P^{(Z)}$ and hence $f^{(Z)}$ depend only on the time to maturity $x = T - t$, i.e. when $P^{(Z)}(t,T,z) = P^{(Z)}(x,z)$ and $f^{(Z)}(t,T,z) = f^{(Z)}(x,z)$, then the no-arbitrage condition reads:

$$
\begin{aligned}
\partial_x f^{(Z)}(x,z) = &\sum_{i=1}^{k} \mu_i^{(Z)}(z)\partial_{z_i} f^{(Z)}(x,z)\\
&+\frac{1}{2}\sum_{i,j=1}^{k} a_{ij}^{(Z)}(z)\Big[\partial^2_{z_i z_j} f^{(Z)}(x,z)\\
&\qquad - \partial_x\left(\int_0^x \partial_{z_i} f^{(Z)}(u,z)du \int_0^x \partial_{z_j} f^{(Z)}(u,z)du\right)\Big]
\end{aligned}
$$

if we use the notation $a^{(Z)}(z) = \sigma^{(Z)}(z)\sigma^{(Z)}(z)^*$. Equivalently,

$$
\begin{aligned}
\partial_x f^{(Z)}(x,z) = &\mu^{(Z)}(z)\cdot\nabla_z f^{(Z)}(x,z)\\
&+\frac{1}{2}\mathrm{trace}(a^{(Z)}(z)\nabla_z^2 f^{(Z)}(x,z))\\
&- \partial_x\left(\int_0^x \nabla_z f^{(Z)}(u,z)du\right) a^{(Z)}(z)\left(\int_0^x \nabla_z f^{(Z)}(u,z)du\right)
\end{aligned}
$$

if we use vector notation. This form of the drift condition implies that if the functions

$$
\partial_{z_i} f^{(Z)}(\,\cdot\,,z),\quad \frac{1}{2}\partial^2_{z_i z_j} f^{(Z)}(x,z) - \partial_{z_i} f^{(Z)}(\,\cdot\,,z)\int_0^{\cdot} \partial_{z_j} f^{(Z)}(u,z)du
$$

$$
1 \le i \le j \le k
$$

are linearly independent, then the drift $\mu^{(Z)}(z)$ and the diffusion $a^{(Z)}(z)$ are determined by the function h. So if the family

$$
\mathcal{H} = \{f^{(Z)}(\,\cdot\,,z)\}_{z\in D}
$$

is used to calibrate the forward curves on a daily basis, then the diffusion process giving the dynamics of the factors Z_t are entirely determined by the parameterization $f^{(Z)}$ of the family of curves!

Notes & Complements

Proposition 2.1 is due to Duffie and Kan, and detailed statements and complete proofs can be found in the original work [50]. Duffie and Kan's paper is one of the seed papers which initiated a wave of publications on affine interest rate models. This culminated with the publication of the paper [49] by Duffie, Filipović and Schachermayer which gives a complete description of the theory of affine Markov processes.

Affine models have been generalized to a wider class of models known as polynomial models of the term structure. The most popular are the quadratic models introduced by El Karoui, Mynemi and Viswanathan in [87], and Jamshidian in [83]. More recently, they were extended by Collin-Dufresne and Goldstein to apply to some HJM and random field models. See [40]. For the record, we mention generalizations to stochastic volatility models as well as hidden Markov models allowing the mean reversion level to jump around in a stochastic fashion. See for example [79]. Finally, Chris Rogers proposed [117] a set of potential models based on a cute idea from the potential theory of Markov processes. Despite their lack of economic foundations, these models lead to easy formulas and numerical implementations.

The stochastic analysis of the stochastic differential equations used in Sect. 2.3.2 can be found in the classical text of Feller [59] or Karatzas and Shreve [86]. More generally, the material covered in the first part of this chapter, including the connection between the lack of arbitrage and the existence of an equivalent martingale measure, is by now classical. It can be found for example in the excellent small book of Lamberton and Lapeyre [97] or in the very well written review article of Björk [8], to which the interested reader is referred to for details and complements. Extra information on mathematical models for the derivative instruments of the fixed income markets can be found in the encyclopedic work of Musiela and Rutkowski [107].

Many authors have tried to reconciliate the finite dimensional nature of the factor models with the potentially infinite dimensional nature of random fields or stochastic dynamics in function spaces as suggested by HJM models. All of these authors rely on one form or another of the PCA tools presented in Chap. 1 to bridge the gap. Galluccio, Guiotto and Roncoronin tried to do just that in a series of two papers [119] and [120] which should be consulted by the reader interested in understanding the dichotomy finite/infinite dimensionality of the mathematical models used in the theory of fixed income markets.

In Sect. 2.4.3 we introduced the notion of maturity-specific risk. This idea is one of the key reasons for the introduction of random field models and HJM models driven by infinite dimensional Wiener processes. However, another notion of maturity-specific risk has been proposed in the literature: for every d date $T_1, \ldots, T_d$, the random vector $(f(t,T_1), \ldots, f(t,T_d))$ has density with respect to the d-dimensional Lebesgue measure. This notion is another way to quantify the idea that the forward curve is a genuinely infinite dimensional object. For instance, an affine model clearly *does not* have this property, but there are finite rank HJM models which do; see the paper of Baudoin and Teichmann [5]. These notions of maturity-specific risk are different and they should not be confused. In this book we are concerned with the *increments* of the rates, rather than the rates themselves.

The idea of modeling the instantaneous forward rates $f(t,T)$ directly as a random field parameterized by two parameters t and T, was first suggested by Kennedy who analyzed in detail the special case of Gaussian fields in [88] and [89]. He considers forward models of the form $f(t,T) = \mu(t,T) + Z(t,T)$ for a given mean zero Gaussian field $\{Z(t,T)\}_{(t,T)\in\mathcal{I}}$ and derives necessary and sufficient conditions on the drift function μ to ensure that the discounted zero coupon bonds are martingales. The random field approach suggested in Sect. 2.4.5 was proposed by Goldstein in [70] where the author generalizes Kennedy's drift condition to this more general setting. It was further generalized by Collin-Dufresne and Goldstein in [40] and Kimmel in [90]. Kimmel's model was chosen by Bester in [6] for a basic simulation model in a numerical comparative study of affine and random field models.

Part II

Infinite Dimensional Stochastic Analysis

3

Infinite Dimensional Integration Theory

We interrupt the flow of the book by breaking away from the interest rate models to start a long excursion in infinite dimensional stochastic analysis. This first chapter of the second part of the book gives a thorough review of the notion of infinite dimensional Gaussian measure, as the latter appears as the most reasonable candidate to support an integration theory in infinite dimensions in view of the absence of analogs of Lebesgue's measure. While preparing for the introduction of Wiener processes and Itô stochastic calculus, we present the various points of view of the cylindrical versus sigma-additive measure controversy, in as agnostic a way as possible.

3.1 Introduction

The factor models introduced earlier to describe the (stochastic) dynamics of the forward curve were driven by a standard finite dimensional Wiener process $W_t = (w_t^{(1)}, \dots, w_t^{(d)})$. It was suggested that investigating the limit as the number d of independent scalar Wiener processes goes to ∞ could bring a solution to some of the shortcomings of the models. In order to implement this idea, we could consider driving the new models by an infinite sequence $W_t = \{w_t^{(j)}\}_{1 \le j < \infty}$ of independent Wiener processes. Treating W_t as an infinite sequence at each time t is the point of view of the theory of cylindrical Wiener process. Instead, we would rather see W_t as a random element of a state space in which the whole stochastic process $\{W_t\}_{t \ge 0}$ could be realized. Obviously, the natural candidate for state space is the space $\mathbb{R}^\infty$ of infinite sequences of real numbers. There are many reasons not to like such a realization of an infinite dimensional Wiener process. Here is a short sample of some of these reasons:

- The space $\mathbb{R}^\infty$ is much too large to be a reasonable state space of an infinite dimensional Wiener process. Most of the space is a wasteland in the sense that it will never be visited by the process. To see why this is indeed the case, recall that, since for each $t > 0$, $\{w_t^{(j)}\}_j$ is an i.i.d. sequence of scalar $N(0,t)$ random variables, one has:

$$\limsup_{j\to\infty} \frac{w_t^{(j)}}{\sqrt{2\log j}} = \sqrt{t} \qquad \text{and} \qquad \liminf_{j\to\infty} \frac{w_t^{(j)}}{\sqrt{2\log j}} = -\sqrt{t}.$$

This shows that, at time t, the random element W_t should *live* in a very small subset of the space of all the sequences, for example the subset of sequences whose large j behavior is given by the two limits above.

- Not only is the size of the space $\mathbb{R}^\infty$ a problem, but its (natural) topology and the corresponding Borel sigma-field are too weak to be amenable to a fine analysis of the process.
- The definition of the d-dimensional Wiener process which we used so far relies on the choice of a coordinate system in $\mathbb{R}^d$. What would happen to the process should we decide to change coordinates? How should we define the limiting process (obtained in the limit $d \to \infty$) in the space $\mathbb{R}^\infty$? We should look for a covariant definition in order to avoid having to rely on coordinate systems.

Different schools of analysts and probabilists have approached the problem differently. We will eventually reconcile the different points of view, but for the time being we make the decision to define infinite dimensional Wiener processes in as intrinsic a manner as possible, and in as tight a state space as possible.

With this in mind, we revisit the definition of a finite dimensional Wiener process which we gave above, and we restate it in a more intrinsic way. A stochastic process $W = \{W_t;\ t \geq 0\}$ is a Wiener process in $E = \mathbb{R}$ or $E = \mathbb{R}^d$ if:

- $W_0 = 0$ almost surely (a.s. for short)
- For each $0 = t_0 < t_1 < \cdots < t_n$ the random variables in E

$$W_{t_n} - W_{t_{n-1}}, \ldots, W_{t_2} - W_{t_1}, W_{t_1} - W_{t_0}$$

  are independent
- For each $0 \leq s < t < \infty$ the distribution of:

$$\frac{1}{\sqrt{t-s}}(W_t - W_s)$$

  is a mean zero Gaussian measure on E which is independent of s and t

Obviously, the notion of Gaussian measure on a finite dimensional Euclidean space is not an issue. In the coordinate version of the definition used above, the measure μ is merely the distribution in $\mathbb{R}^d$ of the d-dimensional Gaussian random vector $(w_1^{(1)}, \ldots, w_1^{(d)})$. It is now clear that, in order to generalize the definition of a Wiener process to a larger class of spaces E (including for example infinite dimensional spaces), we need to define and understand the notion of Gaussian measure on such a space E.

We now explain what we mean by infinite dimensional setting. The discussion above introduced the space $E = \mathbb{R}^\infty$ of infinite sequences of real numbers, and even if we were to settle on such a space as our canonical infinite dimensional setting, we emphasize the need for a clear definition of a topology and a structure of measurable space before a useful notion of E-valued Wiener process can be defined.

3.1.1 The Setting

Even though there are other topological vector spaces of a general type which we will need to use from time to time, we shall try to limit our typical setting to the class of real Banach spaces. We shall try to use the notation E for such a space and E^* for its dual, i.e the space of real-valued continuous linear functions on E. Obviously, the space $E = \mathbb{R}^\infty$ is not a Banach space when equipped with its natural product topology.

The first measure-theoretic concept we shall need is the concept of sigma-field. Because of the topology given by the norm of the Banach space structure, it is natural to consider that E is equipped with its Borel sigma-field $\mathcal{E}$, i.e. the smallest sigma-field containing the open sets. This choice of a sigma-field does guarantee that the continuous functions are measurable. Unfortunately, this sigma-field can be significantly larger than the sigma-field generated by the balls. Since the latter is much easier to deal with when it comes to prove measure theoretic statements, it would be desirable to have both sigma-fields to be identical. This is the case when the Banach space is separable.

So, for the sake of convenience we shall assume that E is separable. This assumption is not restrictive if we limit ourselves to inner regular measures. But most importantly, we shall not lose any generality because most of the classical function spaces are separable. Finally, let us also notice that separability is extremely convenient a feature when present. Indeed, the Borel sigma-field $\mathcal{E}$ is also the sigma-field generated by the balls, or the continuous linear functions on E (i.e. the elements of E^*) or even by any countable set of continuous linear functions dense in a ball of E^*. Dealing with measurability issues will be much easier because of the separability assumption.

The following are typical examples of real separable Banach spaces which we will encounter in the sequel:

- $E = C[0,1]$, the space of continuous real-valued functions on $[0,1]$ equipped with the sup norm $\|f\|_\infty = \sup_{x\in[0,1]} |f(x)|$.
- $E = C_0[0,1]$, the subspace of $C[0,1]$ of the functions f vanishing at 0, i.e. satisfying $f(0) = 0$ (still equipped with the same sup norm).
- $E = H$, a separable real Hilbert space, for example the space $H = L^2(\mathbb{R}, dx)$ of (equivalence classes of) real-valued measurable square-integrable functions on the real line.

We shall also consider Sobolev type spaces that are more regular (i.e. more differentiable functions) as well as weighted spaces for which the norm is computed as a classical norm (such as the sup norm or an L^2-norm) of a multiple of the function, the multiple being given by a weight function having for goal to weight differently the various parts of the domain where the functions are defined. In particular, we shall study a particular weighted Sobolev space H_w in Chap. 6 as a concrete example of a state space for an HJM model.

To illustrate the difficulties which can arise in defining measures on Banach spaces, we note that the unit ball is compact only if the dimension of the space is finite. A consequence of this lack of compactness of the bounded neighborhoods generating the topology of the space is the following *annoying* fact:

Fact. *If E is infinite dimensional, there is* no *sigma-finite translation invariant measure on E.*

In other words, there exists no nontrivial measure μ such that $\mu(A) = \mu(A + x)$ for all $x \in E$ and every A in the Borel sigma-field $\mathcal{E}$. So there is no Haar measure for the additive structure in E, (i.e. no equivalent of the Lebesgue measure) and the theory of integration in E will presumably be more delicate than in the finite dimensional case. In particular, there is no way to define a Gaussian measure via its density. We shall see later in Sect. 3.6 that using densities to define Gaussian measures leads to the notion of cylindrical measure.

3.1.2 Distributions of Gaussian Processes

The classical theory of stochastic processes is a very good source of examples of Gaussian measures in infinite dimensions, namely the distributions of Gaussian processes when viewed as measures on function spaces. We review some of these examples to identify the right abstract definition.

Let us consider for example a real-valued (mean zero) Gaussian process $\xi = \{\xi_t;\, t \in [0,1]\}$ defined on a probability space $(\Omega, \mathcal{F}, \mathbb{P})$. Let $\tilde{E} = \mathbb{R}^{[0,1]}$ be the space of all real-valued functions from $[0,1]$ into $\mathbb{R}$, and let us denote by $\tilde{\mathcal{E}}$ the product sigma-field, generated by cylinder sets of the form

$$\{x \in \tilde{E};\, (x_{t_1}, \dots, x_{t_n}) \in A\}$$

where $t_1, \dots, t_n$ are in $[0,1]$ and $A \in \mathcal{B}_{\mathbb{R}^n}$ the Borel sigma-field of $\mathbb{R}^n$. We can view $\tilde{E} = \mathbb{R}^{[0,1]}$ as a product space of all the real-valued functions on $[0,1]$ and $\tilde{\mathcal{E}}$ is the product sigma-field generated by the cylinders with finite dimensional bases. The coordinate process $\{\tilde{X}_t\}_{t\in[0,1]}$ is defined by $\tilde{X}_t(x) = x(t)$. The coordinate map:

$$X : \Omega \ni \omega \hookrightarrow X(\omega) = \xi_{\cdot}(\omega) \in \tilde{E}$$

is $(\mathcal{F}, \tilde{\mathcal{E}})$ measurable by definition of the product sigma-field $\tilde{\mathcal{E}}$. This map can be used to transport the probability structure given by $\mathbb{P}$ on $(\Omega, \mathcal{F})$ onto a probability structure on $(\tilde{E}, \tilde{\mathcal{E}})$ given by the probability measure $\tilde{\mu}$ defined by:

$$\tilde{\mu}(A) = \mathbb{P}\{\omega \in \Omega;\ X(\omega) \in A\}, \qquad\qquad A \in \tilde{\mathcal{E}}.$$

This probability measure $\tilde{\mu}$ is what is usually called the distribution of the process. But the space $\tilde{E}$ is much too big and its sigma-field $\tilde{\mathcal{E}}$ is too small (the situation is similar but even worse than the one described earlier in the case of the space $\mathbb{R}^\infty$ of countable sequences, i.e. functions on the countable set of integers instead of the continuum $[0,1]$).

In many cases, the auto-covariance function

$$\gamma_\xi(s,t) = \mathbb{E}\{\xi_s \xi_t\}$$

of the process is regular enough for the process ξ to have almost surely continuous sample paths. There exist necessary and sufficient conditions in terms of γ_ξ for this continuity to hold, but we shall not need them here. The interested reader is referred to the Notes & Complements for references. As the sample paths are almost surely continuous, we suspect that the space $\tilde{E}$ of all the real-valued functions on $[0,1]$ could be replaced by the smaller space $E = C[0,1]$ of continuous functions, and that the measure $\tilde{\mu}$ could be replaced by its trace μ on the subset $E = C[0,1]$ of $\tilde{E}$. Indeed, assuming that the sample paths are almost surely continuous should mean that "$\tilde{\mu}(E) = 1$". Unfortunately, E is not a measurable subset of $\tilde{E}$, in the sense that E is not an element of $\tilde{\mathcal{E}}$. Technical measure theoretic manipulations make it possible to get over this obstacle and, because $X(\omega) \in E$ for almost all $\omega \in \Omega$, we can manage to define a measure μ on $\{A \cap E; A \in \tilde{\mathcal{E}}\}$ such that $\mu(E) = 1$. This is our desired measure.

Recall that $E = C[0,1]$ is a real separable Banach space when equipped with the sup norm $\|f\| = \sup_{t\in[0,1]} |f(t)|$. It is easy to see that the Borel sigma-field $\mathcal{E}$ of E is generated by the coordinate maps. We have now a probability measure μ on a real separable Banach space E. But in which sense is this measure Gaussian?

The definition of a Gaussian process states that, for any finite set $\{t_1, t_2, \ldots, t_n\}$ of times in $[0,1]$, the random variables ξ_{t_1}, ξ_{t_2}, $\ldots$ and ξ_{t_n} are jointly Gaussian (i.e. the distribution $\mu_{t_1,t_2,\ldots,t_n}$ of the random vector $(\xi_{t_1}, \xi_{t_2}, \ldots, \xi_{t_n})$ is a Gaussian measure on $\mathbb{R}^n$). Since random variables are jointly Gaussian if and only if any linear combination of these random variables is a scalar Gaussian random variable, we see that for any finite set $\{t_1, t_2, \cdots, t_n\}$ of times in $[0,1]$, and for any finite set $\{a_1, a_2, \cdots, a_n\}$ of real numbers, the random variable

$$f \hookrightarrow \sum_{j=1}^{n} a_j f(t_j)$$

defined on the probability space $(E, \mathcal{E}, \mu)$ is a real-valued Gaussian random variable since its distribution is by definition the distribution of the random variable $\sum_{j=1}^n a_j \xi_{t_j}$ on the original probability space $(\Omega, \mathcal{F}, \mathbb{P})$.

In the present situation, we know everything which needs to be known about the dual space E^* of E, i.e. the space of continuous linear functions on E. Indeed, the Riesz representation theorem states that E^* is the space of signed measures on $[0,1]$ and the duality is given by:

$$\langle \nu, f \rangle = \nu(f) = \int_0^1 f(t)\nu(dt)$$

whenever $\nu \in E^*$ and $f \in E$. Now since any measure on $[0,1]$ appears as the limit of finite linear combination of Dirac delta unit masses at points of $[0,1]$

$$\nu = \sum_{j=1}^{n} a_j \delta_{t_j},$$

where $\delta_t(f) = f(t)$, and since any limit of Gaussian random variables is also a Gaussian random variable, we can conclude that each element ν of the dual E^*, when viewed as a random variable on E via the duality definition:

$$E \ni f \hookrightarrow \langle \nu, f \rangle \in \mathbb{R}$$

is in fact a Gaussian random variable. This property of the distributions of Gaussian processes is what we choose for the definition of a general Gaussian measure in a Banach space.

3.2 Gaussian Measures in Banach Spaces & Examples

As explained above, we assume that E is a real separable Banach space, and we denote by $\mathcal{E}$ its Borel sigma-field, by E^* its dual, and by $\langle x^*, x \rangle = x^*(x)$ the duality between E^* and E given by the evaluation of a continuous linear function on an element of the space.

Definition 3.1. *A probability measure μ on $(E, \mathcal{E})$ is said to be a (mean-zero) Gaussian measure if every $x^* \in E^*$ is a mean-zero real Gaussian random variable $x \hookrightarrow x^*(x)$ on the probability space $(E, \mathcal{E}, \mu)$.*

Example 3.1 (Wiener measure). Because a (scalar) standard Wiener process w is a mean-zero real-valued Gaussian process $w = \{w_t;\ t \in [0,1]\}$ with covariance function $\gamma_w(s,t) = s \wedge t$, and because this covariance function satisfies the conditions for almost sure continuity of the sample paths (see later for details), the derivation given above leads to a Gaussian probability measure μ on the Banach space $C[0,1]$. But since $w_0 = 0$ a.s. the measure μ is in fact concentrated on the subspace $C_0[0,1]$ of functions vanishing for $t = 0$. Since this subspace is closed in $C[0,1]$, and hence is a Borel subset, the measure μ can be viewed as a measure on $E = C_0[0,1]$ equipped with its Borel sigma-field. This measure μ is called the standard Wiener measure.

Example 3.2 (The Ornstein–Uhlenbeck process). Let $w = \{w_t;\ t \in [0,T]\}$ be a standard scalar Wiener process, and let us set:

$$\xi_t = \int_0^t e^{-\beta(t-s)} dw_s$$

for a deterministic constant $\beta > 0$ and all $t > 0$. The process $\xi = \{\xi_t; t \in [0,T]\}$ is a scalar mean zero Gaussian process known as the Ornstein–Uhlenbeck process. It satisfies the linear stochastic differential equation:

$$d\xi_t = -\beta\xi_t dt + dw_t$$

with initial condition $\xi_0 = 0$. We have already come upon this process in our discussion of the Vasicek short-rate model in Chap. 2. Indeed, Vasicek proposed modeling the short rate $\{r_t\}_{t\in[0,T]}$ as the solution of the equation

$$dr_t = (\alpha - \beta r_t)dt + \sigma dw_t.$$

The SDE can be solved explicitly as

$$r_t = e^{-\beta t} r_0 + (1 - e^{-\beta t})\frac{\alpha}{\beta} + \sigma\xi_t$$

where r_0 is the initial short rate. Note that the covariance function

$$\gamma_\xi(s,t) = (2\beta)^{-1}(e^{-\beta|t-s|} - e^{-\beta(t+s)})$$

is again regular enough that the Ornstein–Uhlenbeck process has almost sure continuous sample paths, and that the law of the process ξ can be viewed as a mean-zero Gaussian measure on the Banach space $C[0,T]$. Note that it is in fact supported on the closed subspace $C_0[0,T]$. The law of Vasicek's interest rate process $\{r_t\}_{t\in[0,T]}$ can also be viewed as a Gaussian measure on $C[0,T]$, but this time the mean is not zero but is the given by the deterministic element $t \hookrightarrow e^{-\beta t} r_0 + (1 - e^{-\beta t})\alpha/\beta$ of $C[0,T]$.

We saw that the distributions of (mean-zero) Gaussian processes with continuous sample paths were a good source of examples of (mean-zero) Gaussian measures on Banach spaces. In fact this is the only one. Indeed:

Proposition 3.1. *Any (mean-zero) Gaussian measure μ on a real separable Banach space, say E, is the distribution of a (mean-zero) Gaussian process with continuous sample paths indexed by a compact metric space.*

Proof. Let $U_{E^*} = \{x^* \in E^*;\ \|x^*\| \le 1\}$ be the closed unit ball of the dual space E^*. Equipped with the structure induced by the weak topology of E^*, U_{E^*} is a compact metrizable space, and if for each $x^* \in U_{E^*}$ we define the random variable ξ_{x^*} on the probability space $(E, \mathcal{E}, \mu)$ by $\xi_{x^*} : x \hookrightarrow \langle x^*, x\rangle = x^*(x)$, then μ can be identified with the distribution of the process $\{\xi_{x^*}; x^* \in U_{E^*}\}$ which has continuous (linear) sample paths. □

The procedure described in the proof of the above proposition can be reversed to construct a (mean-zero) Gaussian measure μ on a real separable Banach space E, by constructing a (mean-zero) Gaussian process indexed by the unit ball U_{E^*} and proving that the sample paths of this process are almost surely linear and continuous. Finally, note that since in any Banach space we have

$$\|x\| = \sup_{x^* \in U_{E^*}} x^*(x),$$

the norm appears as the supremum of a Gaussian process.

3.2.1 Integrability Properties

We now assume that we are given a (mean-zero) Gaussian measure μ on a (real separable) Banach space E. Our goal is to analyze the existence of moments for random variables defined on the probability space $(E, \mathcal{E}, \mu)$. If we denote by ψ such a random variable, we are considering the existence of expectations and integrals of the form:

$$\mathbb{E}\{\psi\} = \int_E \psi(x)\, d\mu(x).$$

As in the classical case, sufficient conditions for existence will be derived by comparing ψ to functions of the norm of E, and then by checking the integrability of such functions of the norm. As an example, we may wonder if the integral

$$\int_E \|x\|^2 \mu(dx)$$

is finite. We shall see below that the finiteness of this integral will play a crucial role in the definition of the reproducing kernel Hilbert space of μ. But let us first get some feel for the meaning of the integrability of the norm for the important example of the distribution of a Gaussian process.

Example 3.3. If we assume that μ is the distribution on $E = C[0,1]$ of a mean-zero Gaussian process $\{\xi_t;\, t \in [0,1]\}$ with almost surely continuous sample paths, integrability properties of the norm of E for μ are equivalent to the existence of moments of the supremum of the process ξ, i.e. the random variable $\sup_{0 \le t \le 1} |X_t|$. In this case:

$$\int_E \|x\|^2 \mu(dx) = \int_E \sup_{t\in[0,1]} |x(t)|^2 \mu(dx) = \mathbb{E}\left\{ \sup_{t\in[0,1]} |X_t|^2 \right\}$$

and the problem is to find out when quantities of this type are finite.

These integrability questions have a simple answer in the finite dimensional case. For example, in the case of the real line $E = \mathbb{R}$, not only is the second moment finite, but all moments are finite. In fact much more is known since:

$$\mathbb{E}\{e^{\epsilon X^2}\} = \int_{-\infty}^{+\infty} \frac{1}{\sqrt{2\pi}\sigma} e^{-(\frac{1}{2\sigma^2}-\epsilon)x^2} dx = \frac{1}{\sqrt{1-2\epsilon\sigma^2}}$$

is finite for $\epsilon < \epsilon_0 = (2\sigma^2)^{-1}$. It is natural to ask if a similar result remains true in infinite dimensions. A positive answer was given in 1974 independently by Fernique in a very short note [60] and by Landau and Shepp in a much longer and more technical paper [98]. We state the result below and we reproduce Fernique's elegant proof in one of the Appendices at the end of the chapter.

Theorem 3.1. *If μ is a Gaussian measure on a real separable Banach space E, the integral $\int_E e^{\epsilon\|x\|^2}\mu(dx)$ is finite whenever $\epsilon < \epsilon_0$ where*

$$\epsilon_0 = \left(2 \sup_{x^* \in U_{E^*}} \int_E x^*(x)^2 \mu(dx)\right)^{-1}.$$

Furthermore, the above ϵ_0 is the best possible.

3.2.2 Isonormal Processes

Every Gaussian process we have examined so far has given rise to a Gaussian measure on a Banach space of continuous functions. Indeed, Proposition 3.1 tells us that any continuous Gaussian process on a compact metric space arises in this manner. In this section we examine a very important class of Gaussian processes, the isonormal processes, that are *not* necessarily continuous. They will show up again and again in our study, and play a large role in our presentation of the Malliavin calculus in Chap. 5.

Definition 3.2. *A stochastic process $\{W(h)\}_{h\in H}$ indexed by a Hilbert space H is isonormal (or a white noise) if:*

1. *the random variables $W(h_1), \ldots, W(h_n)$ are jointly mean zero Gaussian for all $h_1, \ldots, h_n$ in H and*
2. *the covariance is given by $\mathbb{E}\{W(g)W(h)\} = \langle g, h\rangle$, where $\langle\cdot,\cdot\rangle : H \times H \to \mathbb{R}$ is the scalar product for H.*

Notice that the definition of an isonormal process prescribes both the mean and the covariance of the process. Since we also assume that it is a Gaussian process, its distribution is completely determined. This uniqueness in distribution makes it possible for us to talk about *the* isonormal process of a Hilbert space, even if we do not always specify the specific probability space on which the process is defined.

The prototypical example of an isonormal process is given by the Wiener integrals $W(h) = \int_0^T h(t)dw_t$ indexed by $h \in L^2([0,T])$, where $\{w_t\}_{t\in[0,T]}$ is a standard scalar Wiener process on a probability space $(\Omega, \mathcal{F}, \mathbb{P})$.

Given a real separable Hilbert space, the isonormal process can easily be constructed by Kolmogorov's extension theorem. However, because of the special structure of Gaussian random variables, a direct construction of the isonormal processes on a separable Hilbert space H can be done as follows: Let $\xi_1, \xi_2, \dots$ be a sequence of independent standard normal random variables on a probability space $(\Omega, \mathcal{F}, \mathbb{P})$, and let $\{e_i\}_i$ be an orthonormal basis of H. Then it is easy to see that the process $\{W(h)\}_{h\in H}$ given by:

$$W(h) = \sum_{i=1}^{\infty} \xi_i \langle e_i, h\rangle$$

is isonormal. The above infinite sum of independent scalar random variables converges almost surely and in any L^p sense because of the three series criterion or a simple martingale argument. Notice also that $h \hookrightarrow W(h)$ is linear and continuous as a map from H into $L^2(\Omega)$ as:

$$\mathbb{E}\{|W(g) - W(h)|^2\} = \mathbb{E}\{W(g-h)^2\} = \|g-h\|^2.$$

One might hope that the isonormal process gives rise to a Gaussian measure on the dual space H^*; unfortunately, this is not the case if H is infinite dimensional. Indeed, since

$$\sup_{\|h\|\le 1} W(h)^2 = \sum_{i=1}^{\infty} \xi_i^2 = +\infty$$

almost surely, it is not possible to define simultaneously all the random variables $W(h)$ outside the same null set, in such a way that the linear map $h \hookrightarrow W(h)$ is continuous on such a common full set. In other words, the infinite series

$$W = \sum_{i=1}^{\infty} \xi_i h_i$$

does not converge in H. See Sect. 3.6.3 below for a modification of this argument leading to the convergence of a similar sum.

We will encounter isonormal processes again in the setting of a Gaussian measure μ on a separable Banach space E, where the role of the indexing Hilbert space H is played by the reproducing kernel Hilbert space H_μ which we now introduce.

3.3 Reproducing Kernel Hilbert Space

It is a well-known fact that the distribution of a (mean-zero) Gaussian process is entirely determined by its covariance function. We exploit this idea in the current abstract framework. The map R defined on E^* by

$$x^* \hookrightarrow R(x^*) = \int_E \langle x^*, x\rangle x \, \mu(dx)$$

will play a crucial role. Note that the integrand $\langle x^*, x\rangle x$ is given by the value of an E-valued random variable defined on the probability space $(E, \mathcal{E}, \mu)$. Indeed, it is the product of the scalar $\langle x^*, x\rangle$ by the element x of E. So this integral needs to be interpreted as the integral of a vector valued function. As the estimate given below shows, this integral is interpreted as a Bochner integral. Some of the standard facts of the integration of vector valued functions are recalled in an appendix at the end of the chapter. The *vector* $R(x^*)$ given by the above integral is a well-defined element of E because of the following estimate:

$$\begin{aligned}\|Rx^*\|_E &\leq \int_E \|x^*(x)x\|_E \;\mu(dx)\\ &= \int_E |x^*(x)|\|x\|_E \;\mu(dx)\\ &\leq \int_E \|x^*\|_{E^*}\|x\|_E^2 \;\mu(dx)\end{aligned}$$

and this last integral is finite because of the integrability Theorem 3.1.

The map $R : E^* \to E$ defined in this way is a bounded linear operator and

$$\|R\| \leq \int_E \|x\|^2 \mu(dx) = C_\mu.$$

Let us consider for a moment the image $R(E^*)$ as a subset of E, and let us define $H = H_\mu$ to be the completion of $R(E^*)$ for the inner product

$$\langle Rx^*, Ry^*\rangle = \int_E x^*(x)y^*(x)\mu(dx).$$

The following question arises: Is H also a subset of E? The answer, fortunately, is yes. Indeed, since the completion can be realized as the set of limits of (equivalent classes of) Cauchy sequences, it is enough to show that sequences in $R(E^*)$ which are Cauchy for the above inner product, do converge to a limit in E. This is indeed the case because, if $\{Rx_n^*\}$ is a Cauchy sequence in $R(E^*)$ for the norm given by the inner product, then:

$$\begin{aligned}\|Rx_n^* - Rx_m^*\|_E = \|R(x_n^* - x_m^*)\|_E &= \left\|\int_E (x_n^* - x_m^*)(x)x\,\mu(dx)\right\|_E\\ &\leq \int_E |(x_n^* - x_m^*)(x)|\;\|x\|_E\mu(dx)\\ &\leq \left(\int_E |(x_n^* - x_m^*)(x)|^2\mu(dx)\right)^{1/2}\left(\int_E \|x\|^2\mu(dx)\right)^{1/2}\\ &= C_\mu^{1/2}\|x_n^* - x_m^*\|_{H^*}\end{aligned}$$

which goes to zero when m and n tend to ∞. In this derivation we used Jensen's and Schwarz's inequalities. Hence, the sequence $\{Rx_n^*\}_n$ is also a Cauchy sequence for the norm of E because the natural injection $i : H \hookrightarrow E$ is continuous. Since E is complete (it is a Banach space after all), this Cauchy sequence converges in E, and in this way, we can identify the completion H with a subset of E. The Hilbert space H_μ so defined is called the reproducing kernel Hilbert space (RKHS for short) of the measure μ.

Since the RKHS plays a such a prominent role in this chapter, we highlight these facts in a definition:

Definition 3.3. *The reproducing kernel Hilbert space (RKHS) H_μ of a Gaussian measure μ on a separable Banach space E is the completion of the image of the map $R : E^* \to E$ defined by*

$$Rx^* = \int_E x^*(x) x \mu(dx)$$

for the norm

$$\|Rx^*\|_{H_\mu} = (x^*(Rx^*))^{1/2}.$$

3.3.1 RKHS of Gaussian Processes

As before we let $\xi = \{\xi_t;\ t \in [0,1]\}$ be a mean-zero real-valued Gaussian process with continuous sample paths, and we let $\gamma(s,t) = \mathbb{E}\{\xi_s \xi_t\}$ be its covariance function. We now construct the RKHS of its distribution. Recall that the latter is a Gaussian measure μ on the Banach space $E = C[0,1]$, and since the dual E^* is the space of signed measures on $[0,1]$, for each fixed $t \in [0,1]$ we have:

$$Rx^*(t) = \delta_t(Rx^*) = \delta_t\left(\int_E x^*(x) x \mu(dx)\right) = \int_E x^*(x) x(t) \mu(dx).$$

if we use the standard notation δ_t for the unit mass at $t \in [0,1]$. Using the fact that each element x^* of the dual space can be identified with a measure on $[0,1]$ with duality between measures and functions given by the integral, i.e. $x^*(x) = \int_{[0,1]} x(s) x^*(ds)$, and using Fubini's theorem we find:

$$\begin{aligned} Rx^*(t) &= \int_E \left(\int_{[0,1]} x(s) x^*(ds)\right) x(t) \mu(dx) \\ &= \int_{[0,1]} \left(\int_E x(s) x(t) \mu(dx)\right) x^*(ds) = \int_{[0,1]} \gamma(s,t) x^*(ds). \end{aligned}$$

Choosing for x^* a finite linear combination of Dirac measures, the above equality gives that for all $\alpha_1, \ldots, \alpha_n$ in $\mathbb{R}$ and for all $t_1, \ldots, t_n$ in $[0,1]$ we have:

$$R(\alpha_1 \delta_{t_1} + \cdots + \alpha_n \delta_{t_n}) = \alpha_1 R\delta_{t_1} + \cdots + \alpha_n R\delta_{t_n} = \alpha_1 \gamma(t_1, \cdot) + \cdots + \alpha_n \gamma(t_n, \cdot)$$

and the inner product between two of these elements is given by:

$$\begin{aligned}\left\langle R\sum_{i=1}^{n}\alpha_i\delta_{t_i}, R\sum_{j=1}^{m}\beta_j\delta_{t_j}\right\rangle &= \sum_{i=1}^{n}\sum_{j=1}^{m}\alpha_i\beta_j\langle R\delta_{t_i}, R\delta_{t_j}\rangle \\ &= \sum_{i=1}^{n}\sum_{j=1}^{m}\alpha_i\beta_j\mathbb{E}\{\xi_{t_i}\xi_{t_j}\} \\ &= \sum_{i=1}^{n}\sum_{j=1}^{m}\alpha_i\beta_j\gamma(t_i,t_j).\end{aligned}$$

Since the set of finite linear combinations of Dirac point masses is dense in the space E^* of signed measures, we can conclude that the notion of RKHS coming out of the abstract construction given in the previous subsection coincides with its classical definition.

3.3.2 The RKHS of the Classical Wiener Measure

Recall that, in the particular case of the standard Wiener process, the covariance is $\gamma(s,t) = s \wedge t$. Let x^* be a measure on $[0,1]$ and

$$\begin{aligned}(Rx^*)(t) = \int_{[0,1]}(s\wedge t)x^*(ds) &= \int_0^t s x^*(ds) + t\int_t^1 x^*(ds) \\ &= \int_0^t x^*([s,1])ds.\end{aligned}$$

So, for the classical Wiener measure, the map R maps the measure x^* into the antiderivative of the complement of the distribution function of the measure x^*. In particular, the function Rx^* is differentiable with explicit derivative $(Rx^*)'(t) = x^*([t,1])$. Computing the inner product reveals:

$$\begin{aligned}\langle Rx^*, Ry^*\rangle &= \int_{[0,1]}\int_{[0,1]}\gamma(s,t)x^*(ds)y^*(dt) \\ &= \int_{[0,1]}\int_{[0,1]}(s\wedge t)x^*(ds)y^*(dt) \\ &= \int_{[0,1]}x^*([t,1])y^*([t,1])dt \\ &= \int_{[0,1]}(Rx^*)'(t)(Ry^*)'(t)dt.\end{aligned}$$

The RKHS of the classical Wiener measure, i.e. the completion of RE^* under this inner product, is the Hilbert space often denoted by $H_0^1[0,1]$ of continuous functions on $[0,1]$ which vanish at zero and which are almost everywhere differentiable, this weak derivative being square-integrable. This space was identified and analyzed in detail by Cameron and Martin in [26] and for this reason, it is usually called the Cameron–Martin space.

3.4 Topological Supports, Carriers, Equivalence and Singularity

Let μ be any probability measure on $(E, \mathcal{E})$. We say that μ is "carried" by $A \in \mathcal{E}$, or that $A \in \mathcal{E}$ is a carrier of μ whenever $\mu(A) = 1$. Carriers are defined up to sets of measure 0. For a given measure, there are many carriers, and there is no canonical way to identify a minimal carrier. We shall not use a special notation for a measure carrier. On the other hand, the topological support (or support for short) will be denoted by $\text{supp}(\mu)$. It is defined as the *smallest closed set* F such that $\mu(F) = 1$, in other words, the smallest carrier among the closed subsets of E.

3.4.1 Topological Supports of Gaussian Measures

It is possible to prove that the topological support of a Gaussian measures is a vector space. In fact, we can be more precise and describe completely the support. Indeed, the latter is nothing but the closure in E of the RKHS H_μ. This is an easy consequence of the Hahn–Banach separation theorem.

Crucial Diagram

Let us assume for a moment that the topological support of μ is the whole space E. In this case, the range $H_\mu = i(H_\mu)$ of the natural inclusion map $i: H_\mu \hookrightarrow E$ is dense, and consequently, its adjoint i^* is also one-to-one with dense range. Thus i^* can be used to identify the dual E^* to the dense subspace $i^*(E^*)$ of the dual H_μ^* of the RKHS H_μ. Hence we have the following diagram:

$$\begin{array}{ccc} E^* & \hookrightarrow & H_\mu^* \\ & & \updownarrow \ (\text{ Riesz identification}) \\ & & H_\mu \hookrightarrow E \end{array}$$

where the one-to-one maps i and i^* can be used to identify H_μ and E^* to dense subspaces of E and H_μ^* respectively, and where R appears once its domain is extended from E^* to H_μ^*, as the Riesz identification of the dual Hilbert space H_μ^* to H_μ. We shall use this diagram frequently in the sequel.

The Hilbert–space structure of the RKHS H_μ determines completely the measure μ, since it contains all the information on the covariance structure of μ. But from the measure theoretical point of view, the set H_μ is negligible. Indeed, $\mu(H_\mu) = 0$ even though:

Proposition 3.2. *If μ is a Gaussian measure on E,*

$$H_\mu = \bigcap_{F vector\ space, \mu(F)=1} F.$$

We already argued that $C_0[0,1]$ was a carrier for the classical Wiener measure. It is in fact the topological support because the classical Cameron–Martin space is dense in $C_0[0,1]$.

The structure captured by the above diagram is known as an abstract Wiener space. It was introduced by L. Gross in 1964. See the Notes & Complements at the end of the chapter for references.

3.4.2 Equivalence and Singularity of Gaussian Measures

Given two measures μ and ν on a measurable space, it is always possible to decompose μ as the sum $\mu = \mu_{ac} + \mu_s$ of a measure μ_{ac} which is absolutely continuous with respect to ν (i.e. a measure given by a density with respect to ν) and a measure μ_s which is singular with respect to ν (i.e. which is carried by a set of ν measure 0). This general decomposition is known under the name of Lebesgue decomposition. We shall see that this decomposition takes a very special form in the case of Gaussian measures. More precisely, if μ and ν are Gaussian, they are either equivalent (absolutely continuous with respect to each other) or singular, in other words, either $\mu = \mu_{ac}$ or $\mu = \mu_s$.

In finite dimension two Gaussian measures are equivalent as long as their topological supports are the same. Indeed, they are equivalent to the Lebesgue's measure of the Euclidean space supporting the two measures. As we are about to see the situation is very different in infinite dimensions.

Two given Gaussian measures are either equivalent or singular, no in-between. This is some surprising form of a 0-1 law: if we consider the Lebesgue's decomposition of one measure with respect to the other one, then the singular part is either zero, or it is equal to the measure itself! Below we give necessary and sufficient conditions for equivalence (and singularity). But the moral of the story should be of the following type:

> Given two generic (mean-zero) Gaussian measures in finite dimension, one can reasonably expect that they will be equivalent. But in infinite dimensional spaces, we should expect them to be singular to each other!

We give the following example as a clear warning that whatever seems natural, should not be taken for granted in infinite dimensions.

Let μ be a mean zero Gaussian measure with topological support the infinite dimensional real separable Banach space E, and for each $t > 0$, let us denote by μ_t the scaled measure defined by $\mu_t(A) = \mu(t^{-1/2}A)$ for all $A \in \mathcal{E}$. Then, all the measures μ_t are singular to each other!

A simple proof can be given using the expansion (3.3) of the next section and the standard law of the iterated logarithm for i.i.d. random sequences. Instead of giving the details, we illustrate this result in the case of the classical Wiener measure. This measure μ is the distribution of the standard (one-dimensional) Wiener process $w = \{w(\tau);\ \tau \in [0,1]\}$. Now, for each $s > 0$ the measure μ_s is the distribution of the process $\sqrt{s}w$. Let us define the set C_s by:

$$C_s = \left\{ x \in C[0,1];\ \limsup_{h \searrow 0} \frac{x(h)}{\sqrt{2h \log|\log h|}} = \sqrt{s} \right\}$$

for $s \geq 0$. The announced singularity follows from the classical law of the iterated logarithm for the standard Wiener process: for every $s \in [0,1]$ we have $\mu_s(C_s) = \mu(C_1) = 1$ yet $\mu_s(C_t) = \mu(C_{t/s}) = 0$ for $s \neq t$.

A precise statement of the dichotomy of Gaussian measures is the following:

Theorem 3.2. *Let E be a real separable Banach space, and let μ and ν be two mean-zero Gaussian measures supporting E with respective RKHS H_μ and H_ν. Then the measures μ and ν are equivalent if and only if:*

- *the subsets H_μ and H_ν of E contain the same elements, and*
- *there exists a self-adjoint Hilbert–Schmidt operator $K \in \mathcal{L}_{\mathrm{HS}}(H_\nu)$ such that*
$$\langle x, y \rangle_{H_\mu} = \langle x, (I+K)y \rangle_{H_\nu}$$
for all $x, y \in H_\mu = H_\nu$.

Throughout the book, we use the notation $\mathcal{L}_{\mathrm{HS}}(H,K)$ for the space of Hilbert–Schmidt operators from a Hilbert space H into another Hilbert space K. Naturally, we use the notation $\mathcal{L}_{\mathrm{HS}}(H)$ when $H = K$. Recall that $A \in \mathcal{L}_{\mathrm{HS}}(H,K)$ if and only if for any CONS $\{e_m\}_m$ and $\{f_n\}_n$ of H and K respectively, we have

$$\sum_{m \geq 1} \|Ae_m\|_K^2 = \sum_{n \geq 1} \|A^* f_n\|_H^2 = \sum_{m,n \geq 1} \langle e_m, f_n \rangle_K^2 = \sum_{m,n \geq 1} \langle e_m, A^* f_n \rangle_H^2 < \infty. \tag{3.1}$$

The space $\mathcal{L}_{\mathrm{HS}}(H,K)$ is always assumed to be equipped with the inner product

$$\langle A, B \rangle_{\mathcal{L}_{\mathrm{HS}}(H,K)} = \mathrm{trace}[AB].$$

Hölder's inequality and the definition (3.1) guarantee the finiteness of this trace. The space $\mathcal{L}_{\mathrm{HS}}(H,K)$ is a (separable whenever H and K are) Hilbert space for this inner product. Notice also that (3.1) implies that $A \in \mathcal{L}_{\mathrm{HS}}(H,K) \Leftrightarrow A^* \in \mathcal{L}_{\mathrm{HS}}(K,H)$ in which case we have the equality of the Hilbert–Schmidt norms $\|A\|_{\mathcal{L}_{\mathrm{HS}}(H,K)} = \|A^*\|_{\mathcal{L}_{\mathrm{HS}}(K,H)}$. For any $e \in H$ and $f \in K$, we use the notation $e \otimes f$ for the rank-one operator from H into K defined by $e \otimes f(h) = \langle e, h \rangle_H f$ for $h \in H$. Condition (3.1) implies that the space of finite linear combinations of these rank-one operators is dense in $\mathcal{L}_{\mathrm{HS}}(H,K)$, and this gives the identification of $\mathcal{L}_{\mathrm{HS}}(H,K)$ and the tensor product $H \otimes K$ of the Hilbert spaces H and K. It is easy to check that, when H is an L^2-space, say $H = L^2(E, \mathcal{E}, \mu)$ for some measure space $(E, \mathcal{E}, \mu)$, then we have the natural identification

$$L^2(E,\mathcal{E},\mu)\otimes K = \mathcal{L}_{\rm HS}(L^2(E,\mathcal{E},\mu),K) = L^2(E,\mathcal{E},\mu:K). \tag{3.2}$$

Here and throughout the book, we use the notation $L^2(E,\mathcal{E},\mu;K)$ or in short $L^2(E;K)$ to denote the space of (μ-equivalent classes) of $\mathcal{E}$-measurable and square integrable functions from E into K, square integrable meaning:

$$\int_E \|\varphi(x)\|_K^2\, d\mu(x) < \infty.$$

3.5 Series Expansions

The goal of this section is twofold. First we want to bridge our treatment of Gaussian measures in Banach spaces with our original introduction of measures on sequence spaces of the form $\mathbb{R}^\infty$. We show that, given the choice of an appropriate basis, the two points of view lead to the same measures. Second, we want to give a computational tool to derive estimates and prove results on the measure theoretic objects constructed from Gaussian measures: series expansions will be our tool of choice to prove these results.

As always in this chapter, H_μ is the RKHS of a Gaussian measure μ on E. Replacing E by a closed subspace if needed, we can always assume that E is the topological support of the measure μ. So without any loss of generality we shall assume that the diagram introduced earlier holds.

Since $i^*(E^*)$ is dense in H_μ^*, it is possible to find a complete orthonormal system (CONS for short) of H_μ^* contained in $i^*(E^*)$ (just apply the Gram–Schmidt orthonormalization procedure to a countable dense set contained in $i^*(E^*)$). So let $\{e_n^*\}_{n\geq 0}$ be a sequence in E^* such that $\{i^*(e_n^*)\}_{n\geq 0}$ is a CONS in H_μ^*. Notice that the sequence $\{e_n\}_{n\geq 0}$ defined by $e_n = Re_n^*$ is the dual CONS system in H_μ, i.e. $\langle e_m^*, e_n\rangle = \delta_{m,n}$ where we use the notation $\delta_{m,n}$ for the Kronecker symbol which is equal to 1 when $m=n$ and 0 otherwise. For the sake of simplifying the notation, we shall identify e_n^* and $i^*(e_n^*)$ on one hand, and e_n and $i(e_n)$ on the other.

On the probability space $(E,\mathcal{E},\mu)$ we define for each integer $n\geq 1$ the map $X_n: E\hookrightarrow H$ by:

$$X_n(x) = \sum_{m=0}^{n} \langle e_m^*, x\rangle e_m.$$

Notice that X_n is defined everywhere on E. From a probabilistic point of view, X_n is a E-valued random variable, while from a functional analysis point of view, X_n is a bounded linear operator from E into E which extends the orthonormal projection of H_μ onto the n-dimensional subspace generated by the first n basis vectors e_m. The following theorem (together with the accompanying remark) is the main result of this section:

Theorem 3.3. *For μ-a.e. $x \in E$,*

$$\lim_{n\to\infty} X_n(x) = x \tag{3.3}$$

the convergence being in the sense of the norm $\|\cdot\|_E$ of E.

Proof. Because of the integrability of Gaussian measures, this result is an immediate consequence of the martingale convergence theorem recalled in the appendix at the end of the chapter. Indeed, if we define the E-valued random variable X on the probability space $(E, \mathcal{E}, \mu)$ by $X(x) = x$, then one can easily check that:

$$X_n = \mathbb{E}\{X|\mathcal{F}_n\}$$

if $\mathcal{F}_n$ denotes the sub sigma-field of $\mathcal{F} = \mathcal{E}$ generated by the (real-valued) random variables $\{e_m^*\}_{0\le m\le n}$. The fact that $\mathcal{F} = \mathcal{E}$ is generated by $\{e_m^*\}_{m\ge 0}$ makes it possible to conclude. □

Remark 3.1. The above result remains true if we start from any CONS $\{e_n^*\}_{n\ge 0}$ in H_μ^*, provided that $\{e_n\}_{n\ge 0}$ is still the dual CONS in H_μ. The only difference is that the E-valued random variables X_n are only defined μ-a.s. instead of being defined everywhere, but the argument of the proof remains the same.

3.6 Cylindrical Measures

Let us assume that μ is a mean-zero Gaussian measure on $(E, \mathcal{E})$ and let us denote by H the reproducing kernel Hilbert space of μ. Without any loss of generality we shall assume that the topological support of μ is equal to the space E.

Given a finite set $\{x_1^*, \ldots, x_n^*\}$ of continuous linear functions in E^* and a Borel subset $A \in \mathcal{B}_{\mathbb{R}^n}$ in the n-dimensional Euclidean space $\mathbb{R}^n$ the set:

$$C(x_1^*, \ldots, x_n^*, A) = \{x \in E;\ (\langle x_1^*, x\rangle, \ldots, \langle x_n^*, x\rangle) \in A\}$$

is called the cylinder with base A and generator $\{x_1^*, \ldots, x_n^*\}$. Cylinders are obviously Borel sets in E. Note the lack of uniqueness: the same cylinder may be obtained from different generators and bases. Their usefulness stands from their simple structure: they are essentially finite dimensional in a (possibly) infinite dimensional space. The set $\mathcal{E}_{\{x_1^*,\ldots,x_n^*\}}$ of cylinders with generator $\{x_1^*, \ldots, x_n^*\}$ is a sigma-field. This sigma-field $\mathcal{E}_{\{x_1^*,\ldots,x_n^*\}}$ can be identified to $\mathcal{B}_{\mathbb{R}^n}$ whenever the elements $x_1^*, \ldots, x_n^*$ of the generator are linearly independent. We shall denote by $\mu_{\{x_1^*,\ldots,x_n^*\}}$ the restriction of μ to the sigma-field $\mathcal{E}_{\{x_1^*,\ldots,x_n^*\}}$. Abusing slightly the notation, we shall sometimes use the same notation for the measure on $\mathbb{R}^n$ given by the identification of $\mathcal{E}_{\{x_1^*,\ldots,x_n^*\}}$ and $\mathcal{B}_{\mathbb{R}^n}$ mentioned above. The net result is that, starting from a (sigma-additive) probability measure μ on $(E, \mathcal{E})$ we end up with a system $\{\mu_{\{x_1^*,\dots,x_n^*\}}\}_{\{x_1^*,\dots,x_n^*\}}$ of finite dimensional measures. Note that these measures satisfy a compatibility condition: if the same cylinder is written using two different generators, the values of the corresponding measures should be the same.

The concept identified in the above discussion is worth a definition.

Definition 3.4. *A cylindrical measure μ on a general topological vector space E is a set $\{\mu_{\{x_1^*,\dots,x_n^*\}}\}_{\{x_1^*,\dots,x_n^*\}}$ of sigma-additive probability measures $\mu_{\{x_1^*,\dots,x_n^*\}}$ on the sigma-fields $\mathcal{E}_{\{x_1^*,\dots,x_n^*\}}$ satisfying the consistency relation.*

Notice that we depart slightly from our habit of considering only real separable Banach spaces. The reason will be made clear below when we discuss Bochner's theorem and the spaces of Schwartz distributions. Because of the consistency hypothesis, a cylindrical (probability) measure μ defines a set function on the set:

$$\mathcal{C} = \bigcup_{\{x_1^*,\dots,x_n^*\}} \mathcal{E}_{\{x_1^*,\dots,x_n^*\}}$$

of all the cylinders by setting:

$$\mu(C) = \mu_{\{x_1^*,\dots,x_n^*\}}(C) \quad \text{whenever} \quad C \in \mathcal{E}_{\{x_1^*,\dots,x_n^*\}}.$$

3.6.1 The Canonical (Gaussian) Cylindrical Measure of a Hilbert Space

We now consider the very particular case of a (real separable) Hilbert space H. First we notice that, because of the Gram–Schmidt orthonormalization procedure, the field $\mathcal{C}$ of cylinders can be defined as the union of the sigma-fields $\mathcal{E}_{\{x_1^*,\dots,x_n^*\}}$ where $\{x_1^*,\dots,x_n^*\}$ is an orthonormal set of vectors in H^*. Then, for each orthonormal set $\{x_1^*,\dots,x_n^*\}$ of vectors in H^*, we define the sigma-additive measure $\mu_{\{x_1^*,\dots,x_n^*\}}$ on the sigma-field $\mathcal{E}_{\{x_1^*,\dots,x_n^*\}}$ by:

$$\mu_{\{x_1^*,\cdots,x_n^*\}}(A) = \frac{1}{\sqrt{(2\pi)^n}} \int_{A_0} e^{-(x_1^2+\cdots+x_n^2)/2} dx_1 \cdots dx_n$$

whenever $A = \{x \in H;\ (\langle x_1^*, x\rangle, \dots, \langle x_n^*, x\rangle) \in A_0\}$ for some Borel subset A_0 of $\mathbb{R}^n$. The cylindrical measure μ defined by this collection $\{\mu_{\{x_1^*,\dots,x_n^*\}}\}$ of probability measures is called the canonical (Gaussian) cylindrical measure of the Hilbert space H. It is the natural infinite dimensional extension of the standard Gaussian measure in finite dimensions.

Unfortunately, the set function μ defined from the prescription of a cylindrical measure in this way is not a bona fide probability measure in general. First $\mathcal{C}$ is not a sigma-field, but worse, μ is most of the time *not* sigma-additive. See the Sect. 3.6.3 below for further discussion of this point.

3.6.2 Integration with Respect to a Cylindrical Measure

Cylindrical measures can be used to integrate cylindrical functions, namely functions of the form $f(x) = \phi(x_1^*(x), \cdots, x_n^*(x))$ for a given function ϕ on $\mathbb{R}^n$ and a set $x_1^*, \cdots, x_n^*$ of continuous linear functions on E. The class of integrable functions can be naturally extended by limiting arguments, and a reasonable integration theory can be developed this way.

One of the main shortcomings of this approach is that it is too often difficult to determine if a given function belongs to the class of integrable functions defined by such a limiting procedure. But most importantly, each time a tool from classical real analysis is needed, a special proof has to be given to make sure that its use is possible in the setting of cylindrical measures.

3.6.3 Characteristic Functions and Bochner's Theorem

The characteristic function of a measure μ on E is the (complex-valued) function on the dual E^* defined by:

$$C_\mu(x^*) = \int_E e^{ix^*(x)}\, d\mu(x), \qquad x^* \in E^*. \tag{3.4}$$

Since functions of a linear form are cylindrical functions, they can be integrated with respect to cylindrical measures. For this reason, the definition given above in formula (3.4) applies as well to cylindrical measures.

Notice that, given $x^* \in E^*$, the map $\lambda \hookrightarrow C_\mu(\lambda x^*)$ is the characteristic function (Fourier transform) of the distribution of x^*, i.e. of the marginal of μ determined by x^*. Consequently, the values of the characteristic function determine completely the one-dimensional marginals, and by linear combinations, all the finite dimensional marginals. In fact, because of the classical Bochner's theorem in finite dimensions, for all practical purposes, the information contained in a characteristic function is the same as the information contained in a cylindrical measure.

From the definition (3.4), it is plain to see that the function C_μ is a non-negative definite function C on E^* which satisfies $C(0) = 1$, and which is weakly continuous at the origin. The converse is true in the sense that all the functions C with these properties are the characteristic functions of cylindrical measures, i.e. are of the form $C = C_\mu$ for some cylindrical measure μ. However, the issue of the sigma-additivity of the measure μ is not settled in general. By definition of a cylindrical measure, the measure μ is (trivially) sigma-additive in finite dimensions. This classical result from harmonic analysis is known as Bochner's theorem. It is the content of Minlos' theorem that this converse is also true for nuclear spaces. In a very informal way, the latter can be characterized as the spaces trying to look like finite dimensional spaces by forcing the bounded neighborhoods of the origin to be compact. The spaces of Schwartz distributions are typical examples of nuclear spaces.

But the converse is not true in general for Hilbert spaces. We already encountered the most famous counterexample to this converse in the case of the canonical cylindrical measure of a Hilbert space. Indeed, if H is a Hilbert space, the function:

$$H^* \ni x^* \hookrightarrow C(x^*) = e^{-\|x^*\|^2}$$

is non-negative definite, weakly continuous at the origin, and satisfies $C(0) = 1$. However, it is not the characteristic function of measure, since the canonical cylindrical measure is not sigma-additive.

3.6.4 Radonification of Cylindrical Measures

Being able to decide whether or not a given cylindrical measure is sigma-additive (and hence can be extended to the sigma-field generated by the cylinder algebra into a bona fide measure) is a difficult problem. This natural question can be generalized in the following form: Given a cylindrical measure μ on a (topological) vector space E and given a (continuous) linear operator A into another (topological) vector space F, which properties of A guarantee that the image cylindrical measure $A\mu$ is sigma-additive on F? This question has a clear answer when E and F are Hilbert spaces.

Theorem 3.4. *If E and F are separable Hilbert spaces, μ is the canonical Gaussian cylindrical measure of E, and A is a bounded linear operator from E into F, then $A\mu$ is sigma-additive if and only if A is a Hilbert–Schmidt operator.*

Proof. Let us fix complete orthonormal systems $\{e_i\}_i$ and $\{f_j\}_j$ in E and F respectively, and in the spirit of our construction of the isonormal process of E given in Sect. 3.2.2, and our discussion of series expansions in Sect. 3.5, we fix a sequence $\{\xi_i\}_i$ of independent standard normal random variables defined on a probability space $(\Omega, \mathcal{F}, \mathbb{P})$. If m and n are integers such that $m < n$,

$$\begin{aligned}
\mathbb{E}\left\{\|\sum_{i=m+1}^{n} \xi_i\, Ae_i\|_F^2\right\} &= \mathbb{E}\left\{\sum_{j=1}^{\infty}\left(\sum_{i=m+1}^{n} \xi_i\,\langle f_j, Ae_i\rangle\right)^2\right\} \\
&= \sum_{j=1}^{\infty}\sum_{i=m+1}^{n} \langle f_j, Ae_i\rangle^2 \\
&= \sum_{i=m+1}^{n} \|Ae_i\|^2
\end{aligned}$$

which shows that

$$\sum_{i=1}^{\infty} \xi_i\, Ae_i \text{ converges in } L^2(\Omega; F) \quad \Longleftrightarrow \quad \sum_{i=1}^{\infty} \|Ae_i\|^2 \text{ converges in } \mathbb{R}. \qquad (3.5)$$

Here we use the notation $L^2(\Omega; F)$ to denote the Hilbert space of F-valued square integrable random variables equipped with the norm

$$\|X\|_{L^2(\Omega;F)} = \mathbb{E}\{\|X\|_F^2\}^{1/2}.$$

In order to prove Theorem 3.4, let us first assume that $A\mu$ is sigma-additive and let us denote by μ_A its sigma-additive extension.

$$\begin{aligned}\int_F \|x\|_F^2 \, d\mu_A(x) &= \int_F \sum_{j=1}^{\infty} \langle f_j, x\rangle^2 \, d\mu_A(x) \\ &= \sum_{j=1}^{\infty} \int_E \langle f_j, Ax\rangle^2 \, d\mu(x) \\ &= \sum_{j=1}^{\infty} \|A^* f_j\|^2\end{aligned}$$

where the integral over E makes sense as the integral of a linear function with respect to a cylindrical measure. So, if the measure μ_A is sigma-additive, Fernique's integrability result implies that that the above quantities are finite, and this proves that the operator A is Hilbert–Schmidt. Conversely, if we assume that A is Hilbert–Schmidt, the equivalence (3.5) shows that the series $\sum_i \xi A e_i$ converges in $L^2(\Omega; F)$, and it is plain to check that the distribution of this sum is a sigma-additive extension of the cylindrical measure $A\mu$. □

3.7 Appendices

Fernique's Proof of the Integrability of Gaussian Measures

Fernique's proof of the existence of exponential moments is based on the following property of (mean zero) Gaussian measures on (measurable) vector spaces. If X and Y are independent random variables with values in a separable Banach space E having the same distribution μ, and if this common law μ is Gaussian, then the E-valued random variables $(X+Y)/\sqrt{2}$ and $(X-Y)/\sqrt{2}$ are independent and also have distribution μ.

In fact, since this property is characteristic of the finite dimensional Gaussian distributions, Fernique uses this property to define Gaussian measures in general (measurable) vector spaces.

The proof is accomplished by finding some positive constants C and ϵ_0 and t_0 such that

$$\mathbb{P}\{\|X\| > t\} \leq C e^{-\epsilon_0 t^2}$$

for all $t > t_0$.

Now, let us notice that, for any positive s and t we have:

$$\begin{aligned}\mathbb{P}\{\|X\| \leq s\}\mathbb{P}\{\|Y\| > t\} &= \mathbb{P}\left\{\frac{\|X-Y\|}{\sqrt{2}} \leq s\right\}\mathbb{P}\left\{\frac{\|X+Y\|}{\sqrt{2}} > t\right\} \\ &= \mathbb{P}\left\{\frac{\|X-Y\|}{\sqrt{2}} \leq s \text{ and } \frac{\|X+Y\|}{\sqrt{2}} > t\right\} \\ &\leq \mathbb{P}\left\{\left|\frac{\|X\|-\|Y\|}{\sqrt{2}}\right| \leq s \text{ and } \frac{\|X\|+\|Y\|}{\sqrt{2}} > t\right\} \\ &\leq \mathbb{P}\left\{\|X\| > \frac{t-s}{\sqrt{2}} \text{ and } \|Y\| > \frac{t-s}{\sqrt{2}}\right\} \\ &= \mathbb{P}\left\{\|X\| > \frac{t-s}{\sqrt{2}}\right\}^2\end{aligned}$$

where we also used the triangle inequality and the fact that X and Y are independent and identically distributed. Letting $t = s+\sqrt{2}t_0$ we can conclude after some rearranging:

$$\frac{\mathbb{P}\{\|X\| > s+\sqrt{2}t_0\}}{\mathbb{P}\{\|X\| \leq s\}} \leq \left(\frac{\mathbb{P}\{\|X\| > t_0\}}{\mathbb{P}\{\|X\| \leq s\}}\right)^2 .$$

Define a sequence $\{t_n\}$ by $t_{n+1} = s + \sqrt{2}t_n$; we see by iteration that

$$\frac{\mathbb{P}\{\|X\| > t_n\}}{\mathbb{P}\{\|X\| \leq s\}} \leq \left(\frac{\mathbb{P}\{\|X\| \leq s\}}{\mathbb{P}\{\|X\| > t_0\}}\right)^{-2^n} .$$

Since by induction $t_n \leq 2^{n/2}(t_0 + 3s)$ then

$$\frac{\mathbb{P}\{\|X\| > t_n\}}{\mathbb{P}\{\|X\| \leq s\}} < \left(\frac{\mathbb{P}\{\|X\| \leq s\}}{\mathbb{P}\{\|X\| > t_0\}}\right)^{-\frac{t_n^2}{(t_0+3s)^2}} .$$

Fix s and t_0 large enough that $\mathbb{P}\{\|X\| \leq s\} > \mathbb{P}\{\|X\| > t_0\}$ and choose

$$\epsilon_0 = \frac{1}{2(t_0+3s)^2}\log\left(\frac{\mathbb{P}\{\|X\| \leq s\}}{\mathbb{P}\{\|X\| > t_0\}}\right)$$

to arrive at the sufficient bound. □

Bochner Integrals

Let $(\Omega, \mathcal{F}, \mu)$ be a measure space, E a separable Banach space, and $\phi : \Omega \hookrightarrow E$ a $(\mathcal{F}, \mathcal{E})$-measurable map. How would one define the integral $\int_\Omega \phi(\omega)\mu(d\omega)$ as an element of E ?

- Let $\phi = \lim_{n\to\infty}^{(E)} \phi_n$, where ϕ_n is simple, that is

$$\phi_n = \sum_{j=1}^{n} x_j \mathbf{1}_{A_j}, \text{ with } x_j \in E \text{ and } A_j \in \mathcal{F}.$$

Notice we have ϕ to be $(\mathcal{F}, \mathcal{E})$-measurable by the separability of E. Define $\int_\Omega \phi(\omega)\mu(d\omega) = \lim_{n\to\infty}^{(E)} \int \phi_n \, d\mu = \lim_{n\to\infty}^{(E)} \sum x_j \mu(A_j)$. This integral, if it exists, is called the "Bochner integral" of ϕ.
A sufficient condition for existence of Bochner integral is

$$\int_\Omega \|\phi(\omega)\|_E \mu(d\omega) < +\infty.$$

In fact, the triangular inequality

$$\left\| \int_\Omega \phi(\omega)\mu(d\omega) \right\|_E \leq \int_\Omega \|\phi(\omega)\|_E \mu(d\omega)$$

is inherited from the construction.

- On the other hand, we could define a vector valued integral in the following way as the unique element $\bar{\mu}_\phi \in E$ such that for all $x^* \in E^*$

$$x^*(\bar{\mu}_\phi) = \int_\Omega x^*(\phi(\omega))\, \mu(d\omega).$$

We need the measurability condition on ϕ: $\forall x^* \in E^*$, $\omega \hookrightarrow x^*(\phi(\omega))$ is measurable. The integral defined this way called the "weak integral." It is this weak integral that will play a central role in constructing generalized bond portfolios in Chap. 6.

Note that if $\bar{\mu}_\phi$ is the Bochner integral of ϕ, then $\bar{\mu}_\phi$ is also the weak integral of ϕ. This follows from the property of the Bochner integral

$$x^*\left(\int_\Omega \phi(\omega)\mu(d\omega)\right) = \int_\Omega x^*(\phi(\omega))\, \mu(d\omega)$$

for all $x^* \in E^*$, which is easily checked by the construction of the integral as a limit of a sum and by the continuity and linearity of each x^*.

Banach Space Valued Martingales

We will make use of the following remarks in the next chapter. Given a probability space $(\Omega, \mathcal{F}, \mathbb{P})$, a filtration $(\mathcal{F}_n)_{n\geq 0}$, and a real separable Banach space E with Borel sigma-field $\mathcal{E}$, a sequence $\{X_n; n \geq 0\}$ of E-valued random variables $X_n : \Omega \to E$ is said to be a (strong or Bochner) martingale if:

1. X_n is $\mathcal{F}_n$-measurable,
2. X_n is integrable,
3. $\mathbb{E}\{X_{n+1}|\mathcal{F}_n\} = X_n$ for all $n \geq 0$.

That is, for each n we have $\mathbb{E}\{\|X_n\|\} < +\infty$ so that expectations can be computed as Bochner integrals. An easy way to understand the notion of conditional expectation used in bullet point 3 is to define $\mathbb{E}\{Y|\mathcal{F}'\}$ as the unique E-valued $\mathcal{F}'$ measurable random variable such that for all $x^* \in E^*$,

$$x^*(\mathbb{E}\{Y|\mathcal{F}'\}) = \mathbb{E}\{x^*(Y)|\mathcal{F}'\}.$$

Note that in the real-valued case, if $X_n \to X_\infty$ a.s and X_∞ is integrable then $X_n = \mathbb{E}\{X_\infty|\mathcal{F}_n\}$ and X_∞ is said to close the martingale.

Theorem 3.5. *Let $\mathcal{F}_\infty = \vee_{n\geq 0}\mathcal{F}_n = \sigma\{\mathcal{F}_n; n \geq 0\}$. If X is an E-valued random variable with $\mathbb{E}\{\|X\|\} < \infty$ then $X_n = \mathbb{E}\{X|\mathcal{F}_n\}$ is a strong martingale and $X_n \to X_\infty = \mathbb{E}\{X|\mathcal{F}_\infty\}$ in $\|\cdot\|_E$ almost surely. In particular, if $\mathcal{F}_\infty = \mathcal{F}$ then $X = X_\infty$.*

Notes & Complements

The interested reader is referred to Bogachev's monograph [18] for an exhaustive presentation of the properties of Gaussian measures in infinite dimensional vector spaces. For all the properties of Gaussian processes, and especially for sufficient conditions for regularity, the reader is referred to the carefully written monograph [1] by Adler where the major contributions of Fernique and Talagrand are reported. More material can also be found in the book [100] by Ledoux and Talagrand.

Integration theory with respect to cylindrical measures was developed in the USSR around Fomin [126], Daletskii [44], and Minlos [105], in France around L. Schwartz and A. Badrikian, [123], and in the US by L. Gross who considered exclusively the case of the canonical cylindrical measure of a Hilbert space, [73] [74]. Gross introduced the notions of measurable norms [73] [74] and abstract Wiener spaces, [74] and proved that this concept was the right framework to *Radonify* the canonical cylindrical measure into a sigma-additive measure on a Banach space. See [74] and the converse due to Satô [121] and Carmona [28]. Theorem 3.4 giving the equivalence between the Radonification of the canonical cylindrical measure in an Hilbert space and the Hilbert–Schmidt property of the Radonifying operator is part of the folklore. Credit for its discovery is shared by different authors in different locations. Using tensor products of Hilbert spaces to represent spaces of Hilbert–Schmidt operators and L^2-spaces of functions with values in a Hilbert space is very convenient. Unfortunately, the definition and the analysis of tensor products of Banach spaces and of more general topological vector spaces are not so simple. We shall discuss briefly the tensor product of covariances in the next chapter without considering these difficulties. Our discussion of the reproducing kernel Hilbert space of a Gaussian measure was inspired by J. Kuelbs' lectures [93]. A stronger form of Proposition 3.2 can be found in [30]. The elementary martingale convergence result given in the appendices above was first proven by Chatterjee.

Stronger results of this type can be found in the book [100] of Ledoux and Talagrand.

In finite dimensions, a sequence of Gaussian measures converges if and only if the sequence of means and the sequence of covariances converges, the respective limits giving the mean and the covariance of the limiting measure. The example of the canonical cylindrical measure of a Hilbert space shows that this result cannot hold in infinite dimensions with the same generality. Nevertheless, a similar result holds when the sequence of *covariance operators* is bounded from above by a covariance operator of a bona fide sigma-additive Gaussian measure. We shall use this result freely in the last chapter of the book. We did not include it in the text because of its technical nature. It was used extensively by Carmona in [28], [29], [3], [34]. The proof of tightness needed for these results can be obtained from an old finite dimensional comparison argument of T.W. Anderson or more modern estimates due to Fernique.

The standard equivalence and singularity dichotomy of Gaussian measures was first proved by Feldman in [58] and Hájek in [78]. See also [75] and [30] for some refinements and consequences relevant to the infinite dimensional Wiener processes discussed in the next chapter.

4

Stochastic Analysis in Infinite Dimensions

The purpose of the previous chapter was more to provide the background for the present chapter than to develop a theory of integration in infinite dimension for its own sake. We now use this preparatory work to present the tools of stochastic analysis which we will need when we come back to interest rate models in the third part of the book. We introduce Wiener processes in infinite dimensions, and we develop a stochastic calculus based on these processes. We illustrate the versatility of this calculus with the solution of stochastic differential equations and the analysis of infinite dimensional Ornstein–Uhlenbeck processes. We then concentrate on Girsanov's theory of change of measure, and on martingale representation theorems, which are two of the most important building blocks of modern continuous time finance.

4.1 Infinite Dimensional Wiener Processes

Given the notion of Gaussian measure in a Banach space introduced in the previous chapter, the analysis of infinite dimensional Wiener processes will be a simple matter of following the strategy outlined at the beginning of the previous chapter.

But before we go parading that we have introduced a new concept, it is important to realize that classical stochastic analysis was already truffled with examples of infinite dimensional Wiener processes. But as *Mister Jourdain* did not know that he was using prose, we may not have been aware of our manipulations of such processes.

4.1.1 Revisiting some Known Two-Parameter Processes

- We first consider the example of the Brownian sheet as introduced by J. Kiefer in the asymptotic analysis of statistical tests of significance. It is the mean-zero Gaussian two-parameter random field $\{\xi(t, \tau);\ t, \tau \in [0, 1]\}$

whose covariance function is given by:

$$\gamma_\xi((t,\tau),(t',\tau')) = \mathbb{E}\{\xi(t,\tau)\xi(t',\tau')\} = (t \wedge t')(\tau \wedge \tau').$$

The two parameters of a Brownian sheet play symmetric roles. If one holds the first parameter fixed, say by setting $t = t_0$ constant, then the process $\xi(t_0,\cdot) = \{\xi(t_0,\tau);\ \tau \in [0,1]\}$ is a Wiener process with variance t_0. Similarly, holding $\tau = \tau_0$ constant yields a scaled Wiener process $\xi(\cdot,\tau_0)$ in the parameter t.

- One could also consider the mean-zero Gaussian random field ξ whose covariance is given by:

$$\gamma_\xi((t,\tau),(t',\tau')) = (t \wedge t')e^{-\lambda|\tau-\tau'|}.$$

In this case, the field is Brownian in t when the parameter τ is held fixed, but it is an *Ornstein–Uhlenbeck* process in τ when the parameter t is held fixed. This process was instrumental in the first developments of the Malliavin calculus.

- More generally, one can consider mean-zero Gaussian processes of the tensor product family for which the covariance γ is obtained from the tensor product of two covariance functions. If $\gamma_1(t,t')$ and $\gamma_2(\tau,\tau')$ are covariance functions for two Gaussian one-parameter processes, then the tensor product

$$\gamma((t,\tau),(t',\tau')) = \gamma_1(t,t')\gamma_2(\tau,\tau')$$

is also a covariance function. In particular, there exists a mean-zero Gaussian two parameter field ξ such that $\gamma_\xi = \gamma$. Moreover many of the regularity properties of the Gaussian processes with covariances γ_1 and γ_2 are inherited by the field ξ.

- Gaussian random fields which are not of a tensor product type have also been analyzed in detail. For instance, as we have seen from Chap. 2, it is fruitful to view the forward rates $\{f(t,T)\}_{0\le t\le T}$ as a continuous two-parameter random field. Kennedy proposed modeling the forward rates by a Gaussian random field with mean $\mu_f(t,T) = \mathbb{E}\{f(t,T)\}$ and covariance $\mathrm{cov}(f(t,T), f(t',T')) = \gamma_f(t,T,t',T')$. The covariance of the field is assumed to have the special structure

$$\gamma_f(t,T,t',T') = c(t \wedge t', T, T')$$

where the function c is symmetric in T and T', and positive in (t,T) and (t',T'). The structure of the covariance is such that the increments of the field are independent, at least when we hold the second parameter (the maturity date) fixed and vary the first parameter (the calendar time):

$$\begin{aligned}
&\mathrm{cov}(f(t_1,T) - f(t_0,T), f(t_2,T) - f(t_1,T)) \\
&\quad = \gamma_f(t_1,T,t_2,T) - \gamma_f(t_0,T,t_2,T) - \gamma_f(t_1,T,t_1,T) + \gamma_f(t_0,T,t_1,T) \\
&\quad = 0.
\end{aligned}$$

The HJM drift condition can then be imposed on the mean via

$$\mu_f(t,T) = \mu_f(0,T) + \int_0^T (c(t \wedge s, s, T) - c(0, s, T))ds.$$

Furthermore, Kennedy proved that if the random field $\{f(t,T)\}_{0 \leq t \leq T}$ is Markovian and stationary, then the form of the covariance is necessarily of the tensor form:

$$c(t,T,T') = e^{-\lambda(t - T\wedge T')} h(|T - T'|)$$

for a real parameter λ. See the Notes & Complements at the end of the chapter for more references.

In both of the first two examples presented above, for each fixed t, the process $\xi(t,\,\cdot\,)$ has almost surely continuous sample paths, so after redefining the process on a set of full probability if necessary, it is possible to view $X_t = \xi(t,\,\cdot\,)$ as an element of the (real separable) Banach space $E = C[0,1]$. Moreover, because of the properties of the covariance kernel $\gamma(t,t') = t \wedge t'$, it is reasonable to envisage the E-valued process $\{X_t;\ t \in [0,1]\}$ as a candidate for the title of Wiener process in E. We shall see below that this guess is well founded.

4.1.2 Banach Space Valued Wiener Process

With the motivation of the discussion of the beginning of the previous chapter, we introduce the following definition for a vector valued Wiener process.

Definition 4.1. *An E-valued process $W = \{W_t, t \geq 0\}$ is called a Wiener process in E if:*

1. $W_0 = 0$
2. *For any finite sequence of times $0 = t_0 < t_1 < \cdots < t_n$, the increments $W_{t_1} - W_{t_0}, \ldots, W_{t_j} - W_{t_{j-1}}, \ldots, W_{t_n} - W_{t_{n-1}}$ are independent E-valued random variables*
3. *There exists a (mean-zero) Gaussian measure μ on E which is the distribution of all the scaled increments $(W_t - W_s)/\sqrt{t-s}$ for $0 \leq s < t < \infty$.*

Obviously, μ is the law of W_1, i.e. $\mu(dx) = \mathbb{P}\{W_1 \in dx\}$, and the distribution of the entire Wiener process $\mathbf{W}$ is completely determined by μ. For this reason, such a process is often called a μ-Wiener process to emphasize the dependence upon the distribution μ of W_1.

4.1.3 Sample Path Regularity

Notice that for $s \neq t$ (we shall assume $s < t$ for the sake of definiteness) we have:

$$\mathbb{E}\{\|W_t - W_s\|_E^p\} = \mathbb{E}\{\|\sqrt{t-s}W_1\|_E^p\} = (t-s)^{p/2} \int_E \|x\|_E^p \mu(dx) \qquad (4.1)$$

where the integral in the right-hand side is finite because of the integrability properties of Gaussian measures. Using Kolmogorov's classical criterion, we see that any Wiener process has a version whose sample paths $[0, \infty) \ni t \to W_t(\omega)$ are continuous for almost all $\omega \in \Omega$. Kolmogorov's criterion was originally proven for real valued stochastic processes, but its proof works as well in the vector valued case, even if the state space is an infinite dimensional Banach space.

As in the real valued case, the proof of Kolmogorov's criterion shows that estimate (4.1) guarantees the almost sure α-Hölder continuity of the sample paths for any $\alpha < 1/2$. But a more precise result holds, and as in the finite dimensional case, the local modulus of continuity is given by the law of the iterated logarithm, and Lévy's uniform modulus of continuity also takes the same form. See the Notes & Complements at the end of the chapter for references.

But one should not think that all the properties of a finite dimensional Wiener process extend to the infinite dimensional setting. The next subsection is intended to help the novice reader avoid sobering experiences with the measure theoretical surprises in infinite dimensions.

4.1.4 Absolute Continuity Issues

Throughout this section we assume that $W = \{W_t, t \geq 0\}$ is a Wiener process in E, and we denote by $\mu_t = \mathcal{L}\{W_t\}$ the distribution of W_t. According to the convention made earlier, we have $\mu = \mu_1$.

The first curiosity which we point out is based on the scaling property $\mu_t(A) = \mu(t^{-1/2}A)$. We saw in the previous chapter that this scaling property implies that all the measures μ_t are singular to each other. More precisely, if E is genuinely of infinite dimension, there exists a family $\{C_t;\, t > 0\}$ of disjoint elements of $\mathcal{E}$ satisfying for every $t > 0$:

$$\mu_t(C_t) = 1 \qquad \text{and} \qquad \mu_s(C_t) = 0 \quad \text{whenever}\, s \neq t.$$

In fact stronger, and more disturbing results hold in the infinite dimensional case. We mention only two of them for the record. The reader interested in the intricacies of these measure theoretic pathologies is referred to the Notes & Complements at the end of this chapter for references.

- There is no sigma-finite measure ν with respect to which all the μ_t are absolutely continuous.
- As a Borel set, the RKHS H is a polar set for the Wiener process: remove this set, and the sample paths of the Wiener process will not be affected! In other words, even though H completely determines μ, and hence all of the μ_t, it is totally irrelevant of the sample paths of the Wiener process W_t. This may sound like a contradiction, but this is a typical instance of the anomalies of infinite dimensional stochastic analysis.

4.1.5 Series Expansions

We now revisit the series expansions derived in the previous chapter. As before we choose a CONS $\{e_n^*, n \geq 1\}$ in H^* made of elements of E^*, and we set $w_n(t) = e_n^*(W_t)$. We first claim that the scalar processes $w_n = \{w_n(t); t \geq 0\}$ are i.i.d. scalar Wiener processes. Indeed, assuming $0 \leq s < t < \infty$ we have:

$$\begin{aligned}
\mathbb{E}\{w_m(s)w_n(t)\} &= \mathbb{E}\{e_m^*(W_s)e_n^*(W_t)\} \\
&= \mathbb{E}\{e_m^*(W_s)e_n^*(W_s)\} + \mathbb{E}\{e_m^*(W_s)e_n^*(W_t - W_s)\} \\
&= s\mathbb{E}\{e_m^*(W_1)e_n^*(W_1)\} \\
&= s\langle e_m^*, e_n^*\rangle_{H^*} \\
&= s\delta_{m,n}
\end{aligned}$$

if we use the independence of the increments and the orthogonality of the e_n^*'s in H_μ^*. Notice that if we set $e_n = Re_n^*$, then $\{e_n; n \geq 1\}$ is the dual orthonormal basis of $\{e_n^*; n \geq 1\}$, and the expansion result of the previous chapter implies that for each fixed time t, we have:

$$W_t = \sum_{n=1}^{\infty} w_n(t)e_n \tag{4.2}$$

where the convergence is almost sure and in the sense of the norm of the Banach space E. In fact, a much stronger result holds. Indeed, for any fixed $T > 0$, the convergence in (4.2) is uniform in $t \in [0, T]$. To see this we apply the same martingale convergence result as before, applying it to the martingale formed by the random elements $X_N = \sum_{n=1}^{N} w(\cdot)e_n$. By the continuity of the sample paths of the scalar Wiener processes, X_N is a random variable in $C([0, T], E)$, the space of E-valued continuous functions on the bounded interval $[0, T]$. This space is a real separable Banach space when equipped with the sup-norm defined by $\|f\|_\infty = \sup_{f\in[0,T]} \|f(t)\|_E$ whenever $f \in C([0, T], E)$. It is then plain to check that $\{X_N; N \geq 1\}$ is a $C([0, T], E)$-valued martingale which converges toward the restriction of W to the interval $[0, T]$.

The series expansion (4.2) will be a very practical tool to manipulate E-valued Wiener processes. It says that, in essence, an infinite dimensional Wiener process can be characterized by a sequence of i.i.d. scalar Wiener processes. This bridges the general theory developed in this chapter with the intuitive discussion from which we started.

Remark 4.1. We revisit the discussion of the introduction of Chap. 3 in which we were led to take the limit $d \to \infty$ and consider an infinite dimensional

Wiener process as an process in the space $\mathbb{R}^\infty$ of sequences of real numbers. So let us suppose that we start with a sequence $\{w_n\}_n$ of i.i.d. scalar Wiener processes; let $e_1 = (1, 0, 0, \ldots)$, $e_2 = (0, 1, 0, \cdots)$ be the natural coordinate elements of $\mathbb{R}^\infty$, and let us formally set $W_t = \sum_{n=1}^\infty w_n(t)e_n = (w_1(t), w_2(t), \ldots)$. Since this W_t could be considered as a good candidate for an infinite dimensional Wiener process, to recast it in the framework presented above we need to answer the following question: in which space E does W_t live? Obvious candidates are:

1. $c_0 = \{x \in \mathbb{R}^\infty; \lim_{n\to\infty} x_n = 0\}$, equipped with the norm $\|x\|_{c_0} = \sup_n |x_n|$.
2. $\ell^2 = \{x \in \mathbb{R}^\infty, \sum x_n^2 = \|x\|_{\ell^2}^2 < \infty\}$ equipped with the inner product

$$\langle x, y\rangle = \sum_n x_n y_n.$$

However, with probability one, $W_t \notin c_0$ and $W_t \notin \ell^2$. Recalling the necessary and sufficient condition for the Radonification of the canonical cylindrical measure in a Hilbert space, we may want to consider enlargements of ℓ^2 obtained by means of weights. More precisely, one may choose a sequence $a = \{a_n\} \in \ell^2$ of square summable weights and define the weighted Hilbert space ℓ_a^2 by:

$$\ell_a^2 = \{x \in \mathbb{R}^\infty; \sum a_n^2 x_n^2 = \|x\|_{\ell_a^2}^2 < \infty\}$$

Then:

Claim. With probability one, $\sum_n a_n^2 w_n(t)^2 < \infty$, and a possible choice of a state space E for the Wiener process is ℓ_a^2.

Notice that for any choice $a \in \ell^2$, we have necessarily $|\sum a_n w_n(t)| < \infty$ almost surely for all t, although $W_t \notin \ell^2$ almost surely! This is very different than the deterministic case: If the deterministic sequence $x = (x_1, x_2, \ldots)$ of real numbers is such that $|\sum a_n x_n| < \infty$ for every $a \in \ell^2$, then necessarily x is an element of ℓ^2.

4.2 Stochastic Integral and Itô Processes

As before, we assume that $W = \{W_t;\, t \geq 0\}$ is a Wiener process defined on a probability space $(\Omega, \mathcal{F}, \mathbb{P})$ and taking values in a real separable Banach space E, and we denote by μ the probability distribution of W_1. On the top of that, we also assume that the probability space is equipped with a filtration $\{\mathcal{F}_t\}_{t\geq 0}$ which satisfies the usual assumptions of right continuity and of saturation of $\mathcal{F}_0$ for the $\mathbb{P}$-null sets, and we assume that W is a $\mathcal{F}_t$-Wiener process in the sense that W_t is $\mathcal{F}_t$ measurable and that for all $0 \leq s < t < \infty$ the increment $W_t - W_s$ is independent of $\mathcal{F}_s$. This is always the case if we work with the filtration $\mathcal{F}_t = \mathcal{F}_t^{(W)}$ generated by the Wiener process, i.e. defined by $\mathcal{F}_t^{(W)} = \mathcal{N} \cup \sigma\{W_s, 0 \le s \le t\}$ if $\mathcal{N}$ denotes the set of $\mathbb{P}$-null sets. We shall also use the notation $\mathcal{P}$ for the predictable sigma-field, i.e. the sigma-field generated by the left continuous processes.

The goal of this section is to make sense of integrals of the form:

$$\int_0^t \sigma_s dW_s \tag{4.3}$$

for integrands $\{\sigma_t;\, t \ge 0\}$ adapted to the filtration. The time-honored approach to the definition of the stochastic integral (4.3) is to first define the integral for simple predictable integrands, and then use a limiting argument to define the integral for a larger class of integrands. A simple predictable integrand is an integrand of the form:

$$\sigma(t,\omega) = \sum_{j=1}^{N} \sigma_j(\omega) \mathbf{1}_{(t_j, t_{j+1}]}(t) \tag{4.4}$$

where $0 = t_1 < t_2 < \cdots < t_{N+1} < \infty$ and each σ_j is $\mathcal{F}_{t_j}$-measurable. For a simple predictable integrand of this form, the stochastic integral (4.3) is naturally defined by the formula:

$$\int_0^t \sigma_s dW_s = \sum_{j=1}^{N} \sigma_j(\omega)(W_{t_{j+1}}(\omega) - W_{t_j}(\omega)). \tag{4.5}$$

Obviously, the nature of the integrand σ has to be such that the *products*

$$\sigma_j(\omega)(W_{t_{j+1}}(\omega) - W_{t_j}(\omega))$$

make sense. Consequently, we distinguish between three types of integrands:

1. The integrand is real valued, in which case the *products* are multiplications of scalars by the increments of the Wiener process and the integral is an element of the state space E of the Wiener process.
2. The integrand is E^*-valued, in which case the *products* are given by the duality between a continuous linear functions on E, say $\sigma_j(\omega)$, and an element of E, say the increment $W_{t_{j+1}}(\omega) - W_{t_j}(\omega)$. In this case, the integral is a real number.
3. The integrand takes values in a space of operators from the space E into another space F which is assumed to be a Hilbert space for the sake of simplicity. In this case, the *products* are given by the evaluation of the operator $\sigma_j(\omega)$, at an element $W_{t_{j+1}}(\omega) - W_{t_j}(\omega)$ of E. In this case, the integral is an element of the space F.

Notice that the second case is merely a particular case of the third one when $F = \mathbb{R}$. But the third case can be derived from the second one because of the following remark. Since an element f of F is entirely determined by the values obtained by computing $\langle f^*, f\rangle$ for all the possible choices of $f^* \in F^*$, one can apply this fact to $f = \int_0^t \sigma_s dW_s$. We get:

$$\begin{aligned}\left\langle f^*, \int_0^t \sigma_s dW_s \right\rangle &= \left\langle f^*, \sum_{j=1}^N \sigma_j (W_{t_{j+1}} - W_{t_j}) \right\rangle \\ &= \sum_{j=1}^N \langle f^*, \sigma_j (W_{t_{j+1}} - W_{t_j})\rangle \\ &= \sum_{j=1}^N \langle \sigma_j(\omega)^* f^*, W_{t_{j+1}} - W_{t_j}\rangle \\ &= \int_0^t \sigma_s^* f^* dW_s.\end{aligned}$$

As before, we use the notation σ^* for the adjoint of the operator σ. In other words, computing f^* of the stochastic integral of the operator valued integrand σ_s is the same thing as integrating the integrand $\sigma_s^* f^*$ which takes values in the dual E^*. We shall use this trick to define the stochastic integral of operator valued integrands, reducing the construction work to the stochastic integral of integrands with values in E^*. Moreover, in order to save time, energy and space, we shall refrain from constructing the stochastic integral from the integral of simple predictable processes. Instead, we rely on the series expansion of the Wiener process to reduce the construction to a classical setting.

As in our discussion of the series expansions of the Wiener process, we choose a CONS $\{e_m^*\}_{m\geq 1}$ of H^* made of elements of the dense subset E^*, and we denote by $\{e_m\}_{m\geq 1}$ the dual basis of H given by the Riesz identification operator $e_m = Re_m^*$. Then, we can expand the Wiener process $\{W_t\}_t$ in a series of the form:

$$W_t = \sum_{m\geq 1} w_m(t) e_m$$

where the $\{w_m\}$ are i.i.d. standard Wiener processes.

4.2.1 The Case of E^* and H^* Valued Integrands

Let us assume that $\lambda = \{\lambda_t;\ t \geq 0\}$ is a predictable process with values in the dual E^*, and let us assume momentarily that it is of the simple predictable type. Then the following derivation is fully justified:

$$
\begin{aligned}
\int_0^t \lambda_s dW_s &= \sum_{j=1}^N \langle \lambda_j, (W_{t_{j+1}} - W_{t_j}) \rangle \\
&= \sum_{j=1}^N \sum_{n \geq 1} \langle \lambda_j, e_n \rangle \langle e_n^*, (W_{t_{j+1}} - W_{t_j}) \rangle \\
&= \sum_{n \geq 1} \sum_{j=1}^N \langle \lambda_j, e_n \rangle (w_{t_{j+1}}^{(n)} - w_{t_j}^{(n)}) \\
&= \sum_{n \geq 1} \int_0^t \langle \lambda_s, e_n \rangle dw_s^{(n)}
\end{aligned}
$$

and moreover:

$$
\begin{aligned}
\mathbb{E}\left\{ \left(\int_0^t \lambda_s dW_s \right)^2 \right\} &= \sum_{n \geq 1} \sum_{m \geq 1} \mathbb{E}\left\{ \left(\int_0^t \langle \lambda_s, e_n \rangle dw_s^{(n)} \right) \left(\int_0^t \langle \lambda_s, e_m \rangle dw_s^{(m)} \right) \right\} \\
&= \sum_{n \geq 1} \mathbb{E}\left\{ \left(\int_0^t \langle \lambda_s, e_n \rangle dw_s^{(n)} \right)^2 \right\} \\
&= \sum_{n \geq 1} \mathbb{E}\left\{ \int_0^t \langle \lambda_s, e_n \rangle^2 ds \right\} \\
&= \mathbb{E}\left\{ \int_0^t \| \lambda_s \|_{H^*}^2 ds \right\}.
\end{aligned}
$$

This last estimate shows that, not only should we be able to integrate adapted processes with values in E^*, but in fact we should be able to integrate all the H^*-valued predictable processes satisfying the energy condition

$$
\mathbb{E}\left\{ \int_0^t \| \lambda_s \|_{H^*}^2 \, ds \right\} < \infty, \qquad \text{for all } t > 0. \tag{4.6}
$$

These elements of H^* are in principle defined only on H and not on E. But because of the interpretation of H^* as a space of (equivalent classes of) random variables, its elements can be defined almost everywhere on E, and the stochastic integral of simple predictable processes with values in H^* still makes sense.

So the final word on the construction of the stochastic integral of vector valued integrands is the following:

If $\lambda = \{\lambda_t;\, t \geq 0\}$ is an adapted process with values in the dual H^* of the RKHS H which satisfies the integrability condition (4.6), then the stochastic integral $\int_0^t \lambda_s dW_s$ exists for every t as a real valued random variable; it satisfies the energy identity:

$$
\mathbb{E}\left\{ \left| \int_0^t \lambda_s dW_s \right|^2 \right\} = \mathbb{E}\left\{ \int_0^t \| \lambda_s \|_{H^*}^2 \, ds \right\} \tag{4.7}
$$

and as a process parameterized by $t > 0$, it is a real valued martingale. Finally, for any CONS $\{e_n^*\}_{n\geq 1}$ of H^*, we have:

$$\int_0^t \lambda_s dW_s = \sum_{n=1}^{\infty} \int_0^t \lambda_s^{(n)} dw_s^{(n)}$$

where the integrands are the predictable processes defined by $\lambda_t^{(n)} = \langle \lambda_t, e_n^* \rangle$ and the integrators are the i.i.d. Wiener processes defined by $w_t^{(n)} = \langle e_n^*, W_t \rangle$.

4.2.2 The Case of Operator Valued Integrands

Let us now assume that F is a real separable Hilbert space and that $\sigma = \{\sigma_t; t \geq 0\}$ is a predictable process with values in the space $L(E, F)$ of bounded linear operators from E into F. A natural definition of the stochastic integral for σ simple predictable suggests that the integral should be an F-valued martingale. If the integral $I_t = \int_0^t \sigma_s dW_s$ exists as an element of F, for any CONS $\{f_n\}_{n\geq 1}$ of F, and if $\{f_n^*\}_{n\geq 1}$ is its dual CONS, one should have:

$$I_t = \sum_{n=1}^{\infty} \langle f_n^*, I_t \rangle f_n \tag{4.8}$$

where the convergence is in the sense of the norm of F, and consequently it is enough to be able to give a meaning to the terms $\langle f_n^*, I_t \rangle$, which according to our previous discussion, should be given by:

$$\langle f_n^*, I_t \rangle = \int_0^t \sigma_s^* f_n^* dW_s = \int_0^t \sigma_s^{(n)} dW_s$$

provided we set $\sigma_s^{(n)} = \sigma_s^* f_n^*$. Notice that since σ_s is a bounded operator from E into F, its transpose σ_s^* is a bounded operator from F^* into E^*, and consequently for each $n \geq 1$, $\sigma^{(n)} = \{\sigma_t^{(n)}; t \geq 0\}$ is a predictable process with values in E^*. Identifying σ_t and $\sigma_t \circ i$ (recall that i is the notation we use for the inclusion of H into E), σ_t can be viewed as an operator from H into F, and as such it is a Hilbert–Schmidt operator since it transforms the canonical cylindrical measure of H into a sigma-additive measure. Since the transpose of a Hilbert–Schmidt operator is also a Hilbert–Schmidt operator, we see that when σ_t^* is viewed as an operator from F^* into H^*, it is a Hilbert–Schmidt operator as well. For the record we give the value of its Hilbert–Schmidt norm $\|\sigma_t^*\|_{\mathcal{L}_{\rm HS}}$ in terms of the CONS $\{f_n^*\}_{n\geq 1}$:

$$\|\sigma_t^*\|_{\mathcal{L}_{\rm HS}}^2 = \sum_{n=1}^{\infty} \|\sigma_t^* f_n^*\|_{H^*}^2. \tag{4.9}$$

Coming back to the definition (4.8) of the stochastic integral via the decomposition into a CONS, we get:

$$\begin{aligned}
\mathbb{E}\left\{\left\|\int_0^t \sigma_s dW_s\right\|_F^2\right\} &= \mathbb{E}\left\{\left\|\sum_{n=1}^{\infty}\langle f_n^*, \int_0^t \sigma_s dW_s\rangle f_n\right\|_F^2\right\} \\
&= \mathbb{E}\left\{\sum_{n=1}^{\infty}\langle f_n^*, \int_0^t \sigma_s dW_s\rangle^2\right\} \\
&= \sum_{n=1}^{\infty}\mathbb{E}\left\{\left(\int_0^t \sigma_s^{(n)} dW_s\right)^2\right\} \\
&= \sum_{n=1}^{\infty}\mathbb{E}\left\{\int_0^t \|\sigma_s^{(n)}\|_{H^*}^2 ds\right\} \\
&= \mathbb{E}\left\{\int_0^t \sum_{n=1}^{\infty} \|\sigma_s^* f_n\|_{H^*}^2 ds\right\} \\
&= \mathbb{E}\left\{\int_0^t \|\sigma_s^*\|_{HS}^2 ds\right\}
\end{aligned}$$

where we used the isometry identity for the stochastic integrals of vector valued integrands, and the definition (4.9) of the Hilbert–Schmidt norm of the transpose of the integrand σ. This computation shows that, not only can we integrate operators from E into a Hilbert space, but in fact we can integrate all the Hilbert–Schmidt operators defined on H, even if we cannot evaluate them on the Wiener process and its increments. Remember, H is polar for the Wiener process!

So the final word on the construction of the stochastic integral of operator valued integrands is the following:

If $\sigma = \{\sigma_t;\, t \geq 0\}$ is a predictable process with values in the space $\mathcal{L}_{\text{HS}}(H, F)$ of Hilbert–Schmidt operators from H into a Hilbert space F which satisfies:

$$\mathbb{E}\left\{\int_0^t \|\sigma_s\|_{\mathcal{L}_{\text{HS}}}^2 ds\right\} < \infty, \qquad \text{for all } t > 0 \tag{4.10}$$

then the stochastic integral $\int_0^t \sigma_s dW_s$ exists for every t as an element of F, it satisfies:

$$\mathbb{E}\left\{\left\|\int_0^t \sigma_s dW_s\right\|_F^2\right\} = \mathbb{E}\left\{\int_0^t \|\sigma_s\|_{\mathcal{L}_{\text{HS}}}^2 ds\right\} \tag{4.11}$$

and as a process parameterized by $t > 0$, it is an F-valued martingale. Finally, for all $f^* \in F^*$ we have:

$$\left\langle f^*, \int_0^t \sigma_s dW_s\right\rangle = \int_0^t \sigma_s^* f^* dW_s.$$

Remarks.

- This last form of the definition of the stochastic integral of operator valued processes corresponds exactly to the definition used for cylindrical Wiener processes.
- We talked freely about operator valued measurability. The following informal definition will be sufficient for the purpose of this book: a map σ with values in a space of operators from E into F is said to be measurable if for all $x \in E$ and $f^* \in F^*$ the real valued function $\langle f^*, \sigma x\rangle$ is measurable.
- By standard localization arguments, the stochastic integral can be extended to (Hilbert–Schmidt) operator valued processes satisfying only:

$$\int_0^t \|\sigma_s\|^2_{\mathcal{L}_{\mathrm{HS}}}\, ds < \infty, \qquad \text{for all } t > 0 \tag{4.12}$$

instead of the stronger (4.10).

4.2.3 Stochastic Convolutions

In many applications, and in particular the study of stochastic partial differential equations, we encounter stochastic integrals of operator valued integrands of the form

$$\int_0^t K(s,t)dW_s,$$

where $\{K(s,t)\}_{0\le s\le t}$ is a Hilbert–Schmidt operator valued process and predictable in the index s. Although the integral is well-defined for each fixed $t \ge 0$, such integrals generally do not define Itô processes.

In this section we study the case of the stochastic convolutions, when the integrand can be factored as

$$K(s,t) = S_{t-s}\sigma_s$$

where for each $t > 0$, S_t is a bounded operator and σ_t a Hilbert–Schmidt operator. One of the reasons for singling out this class of integrands is that the operator valued integrands that arise naturally in the study of HJM models are generally of this form.

Recall that a collection $\{S_t\}_{t\ge 0}$ of bounded linear operators on F is a semigroup if $S_0 = I$ and $S_s \circ S_t = S_{s+t}$ for $s, t \ge 0$. When F is a space of continuous functions on $\mathbb{R}_+$, as in the case of the HJM models studied in Chaps. 6 and 7, the semigroup we deal with most frequently is the semigroup of left shifts defined by

$$(S_t f)(x) = f(t+x).$$

The vector valued process given by the stochastic convolution

$$\int_0^t S_{t-s}\sigma_s dW_s \tag{4.13}$$

is generally not an Itô process. Since the stochastic convolution does not define a martingale in general, the standard martingale inequalities are unavailable. However, the following proposition yields an estimate which we will find useful in proving the existence of mild solutions of SDE in Hilbert spaces.

Proposition 4.1. *Let $\{S_t\}_{t\geq 0}$ be a strongly continuous semigroup of operators on F, and let $\{\sigma_t\}_{t\geq 0}$ be predictable, valued in the space $\mathcal{L}_{\text{HS}}(H,F)$ of Hilbert–Schmidt operators, and satisfying the integrability condition*

$$\mathbb{E}\left\{\int_0^T \|\sigma_t\|^p_{\mathcal{L}_{\text{HS}}(H,F)}\right\} < +\infty$$

for some $p > 2$. Then there exists a constant $C > 0$ such that

$$\mathbb{E}\left\{\sup_{t\in[0,T]}\left\|\int_0^T S_{t-s}\sigma_s dW_s\right\|^p_F\right\} \leq C\mathbb{E}\left\{\int_0^T \|\sigma_t\|^p_{\mathcal{L}_{\text{HS}}(H,F)}\right\}.$$

Proof. The proof of the proposition is based on the factorization method introduced by Da Prato, Kwapien, and Zabczyk. See [43] for details. First we note that for every $0 < \alpha < 1$ and $0 < u < t$ we have the beta integral

$$\int_u^t (t-s)^{\alpha-1}(s-u)^{-\alpha} ds = \frac{\pi}{\sin \pi\alpha}.$$

Letting

$$Y_s = \int_0^s (t-s)^{-2/(p+2)} S_{s-u}\sigma_u dW_u$$

we have by the stochastic Fubini theorem and the above identity with $\alpha = 2/(p+2)$ that the stochastic integral can be written as a Bochner integral

$$\int_0^t S_{t-u}\sigma_u dW_u = \frac{\sin\left(\frac{2\pi}{p+2}\right)}{\pi}\int_0^t (t-s)^{-p/(p+2)} S_{t-s} Y_s ds.$$

Hence by Hölder's inequality we have

$$\begin{aligned}
\sup_{t\in[0,T]}\left\|\int_0^t S_{t-u}\sigma_u dW_u\right\|^p &\leq \pi^{-p}\sup_{t\in[0,T]}\left\|\int_0^t (t-s)^{-p/(p+2)} S_{t-s}Y_s ds\right\|^p \\
&\leq \pi^{-p}M^p\left(\int_0^T s^{-p^2/((p+2)(p-1))} ds\right)^{(p-1)/p}\int_0^T \|Y_s\|^p ds \\
&\leq \pi^{-p}M^p\left(\frac{2p^2}{p-2}\right)^{p-1} T^{(p-2)/(p+2)}\int_0^T \|Y_s\|^p ds
\end{aligned}$$

where $M = \sup_{t\in[0,T]} \|S_t\|$. Now the following bound

$$\mathbb{E}\{\|Y_s\|^p ds\} = \mathbb{E}\left\{\left\|\int_0^s (s-u)^{-2/(p+2)} S_{s-u}\sigma_u dW_u\right\|^p\right\}$$
$$\leq C_p M^p \mathbb{E}\left\{\left(\int_0^s (s-u)^{-4/(p+2)} \|\sigma_u\|^2_{\mathcal{L}_{\text{HS}}(G,F)} du\right)^{p/2}\right\}$$

follows from a moment inequality for stochastic integrals, with $C_p \leq p^p 2^{p^2/2-p}$. Finally, Young's inequality implies

$$\int_0^T \left(\int_0^s (s-u)^{-4/(p+2)} \|\sigma_u\|^2_{\mathcal{L}_{\text{HS}}(G,F)} du\right)^{p/2} ds$$
$$\leq \int_0^T \|\sigma_u\|^p_{\mathcal{L}_{\text{HS}}(G,F)} du \left(\int_0^T u^{-4/(p+2)} du\right)^{p/2}$$
$$= \left(\frac{p+2}{p-2}\right)^{p/2} T^{p(p+2)/(2(p-2))} \left\{\int_0^T \|\sigma_u\|^p_{\mathcal{L}_{\text{HS}}(G,F)} du\right\}.$$

Choosing a constant C satisfying

$$C > \pi^{-p} M^{2p} T^{p/2-1} 2^{p^2} p^{7p/2} (p-2)^{-(3p/2-1)}$$

completes the proof. □

4.3 Martingale Representation Theorems

Throughout this section we assume that the filtration is generated by the Wiener process, i.e. that $\mathcal{F}_t = \mathcal{F}_t^{(W)} = \mathcal{N} \cup \sigma\{W_s;\ 0 \leq s \leq t\}$ where as usual $\mathcal{N}$ denotes the field of the sets of measure zero. That is, up to sets of zero probability, for each t, the sigma-field $\mathcal{F}_t$ is the smallest sigma-field containing the sets of the form $\{W_s \in A\}$ for $A \in \mathcal{E}$, and $0 \leq s \leq t$. The fact that E is separable implies that the Borel sigma-field $\mathcal{E}$ of E is generated by the elements of the dual E^*, but also by any (countable) subset of E^* as long as this set is total for the weak topology. In particular, this implies that:

$$\begin{aligned}
\mathcal{F}_t^{(W)} &= \mathcal{N} \cup \sigma\{x^*(W_s);\ x^* \in E^*, 0 \leq s \leq t\} \\
&= \mathcal{N} \cup \sigma\{x^*(W_s);\ x^* \in H^*, 0 \leq s \leq t\} \\
&= \mathcal{N} \cup \sigma\{w_n(s);\ n \geq 1, 0 \leq s \leq t\}
\end{aligned}$$

where as before, the i.i.d. (scalar) Wiener processes $w^{(n)}$ are obtained from a CONS of H^*.

The purpose of this section is to prove a representation theorem for martingales in the filtration of the Wiener process. This theorem is not much different from the corresponding finite dimensional analog. We give a complete proof for the sake of completeness.

Theorem 4.1 (Martingale Representation Theorem). *For every square-integrable martingale $M = \{M_t\}_{t\geq 0}$ with values in a real separable Hilbert space F, there exists a square-integrable $\sigma = \{\sigma_t\}_{t\geq 0}$ predictable process with values in $\mathcal{L}_{\mathrm{HS}}(H, F)$ such that*

$$M_t = M_0 + \int_0^t \sigma_s dW_s, \qquad t \geq 0. \tag{4.14}$$

Proof. Without any loss of generality we can assume that $M_0 = 0$. Moreover, because of the right continuity of the filtration, we can reduce the proof to proving that, for each $T > 0$ and each square-integrable F-valued $\mathcal{F}_T$-measurable random variable M_T, there exists a square-integrable $\sigma = \{\sigma_t\}_{0\leq t\leq T}$ predictable process with values in $\mathcal{L}_{\mathrm{HS}}(H, F)$ such that (4.14) holds with $t = T$. Finally, the general case of a Hilbert space F can be reduced to the particular case of $F = \mathbb{R}$ by decomposing M_t on a CONS of F, and reconstructing the operator σ_t from its one-dimensional projections on the basis vector. We do not give the details of this last step because we shall only need the martingale representation result in the case $F = \mathbb{R}$. So we consider that $T > 0$ is fixed, we define the set $\mathcal{I}_T$ by:

$$\mathcal{I}_T = \left\{ \int_0^T \lambda_s dW_s;\ \lambda = \{\lambda_t\}_{0\leq t\leq T} \text{ square-integrable, predictable in } H^* \right\},$$

and we prove that $L_0^2(\Omega, \mathcal{F}_T, \mathbb{P}) = \mathcal{I}_T$. where $L_0^2(\Omega, \mathcal{F}_T, \mathbb{P})$ is the subspace of mean-zero random variables in $L^2(\Omega, \mathcal{F}_T, \mathbb{P})$. Since $\mathcal{I}_T$ is closed in L^2 because of the energy identity (4.7), it is enough to show that:

$$\xi \in L_0^2(\Omega, \mathcal{F}_T, \mathbb{P}) \text{ and } \xi \perp \mathcal{I}_T \text{ imply } \xi = 0.$$

Now let $\lambda \in \mathcal{I}_T$ be deterministic and bounded, and let us set:

$$N_t = \int_0^t \lambda_s dW_s = \sum_{n\geq 1} \int_0^t \lambda_s^{(n)} dw_s^{(n)}$$

where as usual, $\lambda_s^{(n)} = \langle \lambda_s, e_n \rangle$, and $w_s^{(n)} = e_n^*(W_s)$. $\{N_t;\ 0 \leq t \leq T\}$ is a square-integrable continuous martingale. Let $\{M_t;\ 0 \leq t \leq T\}$ be the solution of the equation:

$$dM_t = M_t dN_t$$

with initial condition $M_0 = 1$. We know that the solution of this equation is given by:

$$M_t = \exp\left(N_t - \frac{1}{2} \ll N, N \gg_t\right) = \exp\left(\int_0^t \lambda_s dW_s - \frac{1}{2}\int_0^t \|\lambda_s\|_{H^*}^2 ds\right),$$

where $\ll \cdot, \cdot \gg$ denotes the quadratic variation between semi-martingales. Since λ is bounded, the random variable M_t is obviously square-integrable, and moreover:

$$M_T = C^{-1} \exp\left(\int_0^T \lambda_s dW_s\right)$$

for some deterministic constant C. Now

$$\begin{aligned} \mathbb{E}\left\{\xi \exp\left(\int_0^T \lambda_s dW_s\right)\right\} &= C\mathbb{E}\{\xi M_T\} \\ &= C\mathbb{E}\left\{\xi\left(1 + \int_0^T M_s \lambda_s dW_s\right)\right\} \\ &= 0 \end{aligned}$$

because the process $\{M_t\lambda_t\}_{0\le t\le T}$ is in $\mathcal{I}_T$. For each integer $n \ge 1$, for each subset $\{t_1, \dots, t_n\}$ of $[0, T]$, for each subset $\{\gamma_1, \dots, \gamma_n\}$ of real numbers, and for each subset $\{x_1^*, \dots, x_n^*\}$ of $E^* \subset H^*$ we conclude that:

$$\mathbb{E}\left\{\xi \exp\left(\sum_{j=1}^n \gamma_j x_j^*(W_{t_j})\right)\right\} = 0 \tag{4.15}$$

by applying the previous result to:

$$\lambda_t = \sum_{j=1}^n \gamma_j \mathbf{1}_{[0,t_j]}(t) x_j^*.$$

Since the real valued random variables $x_j^*(W_{t_j})$ generate (together with the null sets) the sigma-field $\mathcal{F}_T$, (4.15) implies that $\xi = 0$. □

Remark 4.2. If $M = \{M_t\}_t$ is only a local martingale then a standard localization argument can be used to prove that there exists a predictable H^*-valued process $\lambda = \{\lambda_t\}_t$ satisfying:

$$\int_0^t \|\lambda_s\|_{H^*}^2 ds < \infty \qquad \text{for all } t > 0 \tag{4.16}$$

and such that Eq. (4.14) holds. Remember the condition (4.12) under which the stochastic integral can still make sense. In particular, any local martingale in this filtration is continuous (since it is a stochastic integral).

4.4 Girsanov's Theorem and Changes of Measures

Let $\lambda = \{\lambda_t;\, t \ge 0\}$ be a predictable process with values in the dual H^* of the RKHS H, and let us assume as before that:

$$\mathbb{E}\left\{\int_0^t \|\lambda_s\|_{H^*}^2\, ds\right\} < \infty, \qquad \text{for all } t > 0. \tag{4.17}$$

Then we can consider the (real valued) square-integrable martingale $M = \{M_t\}_{t\ge 0}$ defined by the stochastic integral $M_t = \int_0^t \lambda_s dW_s$. Its quadratic variation process is $\ll M \gg_t = \int_0^t \|\lambda_s\|_{H^*}^2 ds$ and consequently, the process $Z = \{Z_t\}_{t\ge 0}$ defined by:

$$Z_t = e^{M_t - \frac{1}{2}\ll M,M\gg_t} = \exp\left(\int_0^t \lambda_s dW_s - \frac{1}{2}\int_0^t \|\lambda_s\|_{H^*}^2 ds\right) \tag{4.18}$$

is a non-negative local martingale. The process $\{Z_t\}_{t\ge 0}$ is usually called the Doleans exponential of the (local) martingale M_t. We shall assume that it is in fact a martingale (i.e. that $\mathbb{E}\{Z_T\} = 1$), allowing us to define a probability measure $\mathbb{Q}$ by its restrictions to all the sigma-fields $\mathcal{F}_t$, these restrictions being given by their densities with respect to the corresponding restrictions of $\mathbb{P}$. More precisely, we define $\mathbb{Q}$ so that:

$$\left.\frac{d\mathbb{Q}}{d\mathbb{P}}\right|_{\mathcal{F}_t} = \exp\left(\int_0^t \lambda_s dW_s - \frac{1}{2}\int_0^t \|\lambda_s\|_{H^*}^2 ds\right). \tag{4.19}$$

Sufficient conditions for Z_t to be a martingale exist. The most well known of them is presumably Novikov's condition. Note that we implicitly assumed that $\mathcal{F}$ was the smallest sigma-field containing all the $\mathcal{F}_t$'s. Our version of the famous Girsanov theorem is the following:

Theorem 4.2 (Girsanov's Theorem). *The E-valued process $\tilde{W} = \{\tilde{W}_t;\, t \ge 0\}$ defined by:*

$$\tilde{W}_t = W_t - \int_0^t R\lambda_s\, ds \tag{4.20}$$

is a Wiener process for the measure $\mathbb{Q}$, i.e. on the probability structure $(\Omega, \mathcal{F}, \{\mathcal{F}\}_t)_{t\ge 0}, \mathbb{Q})$, where the linear map R is the Riesz identification of H^* onto H.

Proof. Since $\{\tilde{W}_t\}_t$ is a bona fide E-valued process (i.e. not cylindrical), we only need to prove that for each $x^* \in E^*$, the process $\{x^*(\tilde{W}_t)\}_t$ is a scalar $\mathcal{F}_t$-Wiener process the variance of which is the square of the norm of x^* in the dual of the RKHS of $\tilde{W}$. In other words, it is enough to prove that for each $x^* \in E^*$,

$$\exp\left(x^*(\tilde{W}_t) - t\|x^*\|_{H^*}^2/2\right)$$

is an $\mathcal{F}_t$-local martingale for $\mathbb{Q}$. Because of the characterization of $\mathbb{Q}$-martingales in terms of $\mathbb{P}$-martingales given above, this is equivalent to proving that

$$\exp\left(x^*(\tilde{W}_t) - t\|x^*\|^2_{H^*}/2\right) Z_t$$

is an $\mathcal{F}_t$-local martingale for $\mathbb{P}$. But given the definition of the density process Z_t and the fact that:

$$x^*(\tilde{W}_t) = x^*(W_t) - \int_0^t x^*(R\lambda_s)ds = x^*(W_t) - \int_0^t \langle x^*, \lambda_s\rangle_{H^*} ds,$$

this is an easy consequence of the fact that $\{\int_0^t (x^*\lambda_s)dW_s\}_{t\geq 0}$ is a continuous $\{\mathcal{F}_t\}_{t\geq 0}$-martingale for $\mathbb{P}$ with quadratic variation

$$t\|x^*\|^2_{H^*} - 2\int_0^t \langle x^*, \lambda^*_s\rangle_{H^*} ds + \int_0^t \|\lambda_s\|^2_{H^*} ds.$$

The proof is now complete. □

Changes of Measures

The above result says that, if we start from an E-valued Wiener process W on the probability structure $(\Omega, \mathcal{F}, (\mathcal{F}_t)_{t\geq 0}, \mathbb{P})$, for each H^*-valued integrand λ for which the Doleans exponential of the stochastic integral $\int_0^t \lambda_s dW_s$ is a bona fide martingale, one can define a probability measure $\mathbb{Q}$ which is locally equivalent to $\mathbb{P}$, and we can even recover an E-valued Wiener process $\tilde{W}$ by correcting W_t (which is no longer a Wiener process on the probability structure induced by $\mathbb{Q}$) with a bounded variation drift constructed from λ.

In many financial applications, we start with a probability measure $\mathbb{P}$ (called the historical measure) and in order to compute prices by expectations, we replace $\mathbb{P}$ by an equivalent measure with turns the discounted prices of all the tradable instruments into martingales (a so-called equivalent martingale measure). In order to prepare for this passage to the risk neutral world, we first investigate the class of all the equivalent probability measures. When we say equivalent, we really mean locally equivalent in the sense that for each $t > 0$, the restriction of $\mathbb{Q}$ to $\mathcal{F}_t$, say $\mathbb{Q}_t$, is equivalent to the restriction of $\mathbb{P}$ to $\mathcal{F}_t$, say $\mathbb{P}_t$, i.e. $\mathbb{Q}_t \sim \mathbb{P}_t$ for all t. So what can we say about such measures $\mathbb{Q}$?

If $\mathbb{Q}$ is locally equivalent to $\mathbb{P}$, let Z_t be the density (i.e. the Radon–Nykodym derivative) of $\mathbb{Q}_t$ with respect to $\mathbb{P}_t$. It is easy to see that $Z = \{Z_t;\, t \geq 0\}$ is a non-negative $\mathbb{P}$-martingale with expectation 1. The filtration $\{\mathcal{F}_t\}_t$ being assumed to be right continuous, the sample paths of the martingale $\{Z_t\}_t$ can be assumed to be almost surely right continuous.

An equivalent form of the definition of the martingale Z is that for every $0 \le s < t$, and every $\mathcal{F}_t$-measurable random variable X we have:

$$\mathbb{E}^{\mathbb{Q}}\{X|\mathcal{F}_s\} = \frac{1}{Z_s}\mathbb{E}^{\mathbb{P}}\{XZ_t|\mathcal{F}_s\}$$

or even that an adapted and $\mathbb{Q}$-integrable process $\{X_t\}_t$ is a $\mathbb{Q}$-martingale if and only if the process $\{X_t Z_t\}_t$ is a $\mathbb{P}$-martingale.

If we now assume that the filtration $\{\mathcal{F}\}_t$ is the filtration $\{\mathcal{F}_t^{(W)}\}_t$ of an E-valued Wiener process, then $\{Z_t\}_t$ is a local martingale with continuous sample paths, and applying Itô's formula to compute $\log Z_t$, we see that:

$$Z_t = \exp\left(M_t - \frac{1}{2} \ll M, M \gg_t\right)$$

for some (continuous) local martingale $\{M_t\}_t$. Applying the martingale representation theorem to this local martingale, we get the existence of a predictable H^*-valued process $\{\lambda_t\}_t$ such that (4.16) and $M_t = \int_0^t \lambda_s dW_s$. So we have completed the journey, we are back where we started from, in the situation described in the statement of the Girsanov's theorem for which:

$$Z_t = \exp\left(\int_0^t \lambda_s dW_s - \frac{1}{2}\int_0^t \|\lambda_s\|_{H^*}^2 ds\right).$$

4.5 Infinite Dimensional Ornstein–Uhlenbeck Processes

In this section, we review various facts about Ornstein–Uhlenbeck (OU) processes, and especially their infinite dimensional versions. Some of the results we review will be used in the study of the specific interest rate models we consider in Chap. 7. Moreover, the intuition which we develop will help us extend some of the theoretical results to situations of interest.

4.5.1 Finite Dimensional OU Processes

A one-dimensional OU process is usually defined as the solution of the stochastic differential equation:

$$d\xi_t = -a\xi_t dt + b dw_t \tag{4.21}$$

where the a and b are real numbers and where $\{w_t\}_{t\ge 0}$ is a one-dimensional Wiener process. The coefficient a appearing in the drift is often assumed to be positive, for when this is the case, the drift is a restoring force reverting ξ_t toward its mean. The initial condition, say ξ_0, is usually (implicitly) assumed to be independent of the driving Wiener process. The solution can be written explicitly in the form:

$$\xi_t = e^{-ta}\xi_0 + b\int_0^t e^{-(t-s)a}dw_s, \qquad 0 \le t < +\infty. \tag{4.22}$$

The mean and the covariance of the solution are given by the formulas:

$$\mathbb{E}\{\xi_t\} = e^{-ta}\mathbb{E}\{\xi_0\} \tag{4.23}$$

and

$$\operatorname{cov}\{\xi_s, \xi_t\} = \left[\operatorname{var}\{\xi_0\} + \frac{b^2}{2a}(e^{2(s\wedge t)a} - 1)\right] e^{-(s+t)a}. \tag{4.24}$$

Notice that $\{\xi_t\}_t$ is a Gaussian process whenever ξ_0 is Gaussian. The solution process is a Markov process. It has a unique invariant measure, the normal distribution $N(0, b^2/2a)$. Starting with this distribution, the Markov process becomes a stationary ergodic Markov (mean-zero) Gaussian process. Its covariance is then given by the formula:

$$\mathbb{E}\{\xi_s\xi_t\} = \frac{b^2}{2a}e^{-a|t-s|}. \tag{4.25}$$

Let us now review the concept of an n-dimensional (n finite) OU process, say $\{X_t;\ t \ge 0\}$. As we can see by comparing formulas (4.21–4.25) to formulas (4.26–4.30) below, in order to go from the definition of a one-dimensional OU process to the definition of the n-dimensional analog we need the following substitutions:

$-a \in \mathbb{R}$	$\longrightarrow$	A $n \times n$ matrix
$b \in \mathbb{R}$	$\longrightarrow$	B $n \times n$ matrix
$\{w_t\}_{t\ge 0}$ $1-$ dim. Wiener Process	$\longrightarrow$	$\{W_t\}_{t\ge 0}$ $n-$ dim. Wiener Process

In this way, an n-dimensional OU process can be defined as the solution of a system of stochastic equations written in a vector form as:

$$dX_t = AX_t dt + BdW_t \tag{4.26}$$

where $\{W_t\}_{t\ge 0}$ is an n-dimensional Wiener process. Note that B really does not need to be a square matrix. It could be an $n \times d$ matrix as long as the Wiener process W is d-dimensional.

As before, the solution of the stochastic system can be written explicitly:

$$X_t = e^{tA}X_0 + \int_0^t e^{(t-s)A}BdW_s, \qquad 0 \le t < +\infty. \tag{4.27}$$

If X_0 is assumed to be independent of the driving Wiener process, the mean and the covariance are given by the formulas:

$$\mathbb{E}\{X_t\} = e^{tA}\mathbb{E}\{X_0\} \tag{4.28}$$

and

$$\mathrm{cov}\{X_s, X_t\} = e^{sA}\mathrm{var}\{X_0\}e^{tA^*} + \int_0^{s\wedge t} e^{(s-u)A}BB^*e^{(t-u)A^*}\,du. \qquad (4.29)$$

As before, $\{X_t\}_t$ is a Gaussian process as soon as the initial condition X_0 is Gaussian. We shall always restrict ourself to this case. The process is a strong Markov process.

In the same way a is often assumed to be positive to guarantee the ergodicity of the process and the convergence as $t \to \infty$ toward an invariant measure, the drift matrix A is often assumed to be strictly dissipative in the sense that

$$\langle Ax, x\rangle < 0 \text{ for all } x \in \mathbb{R}^n,\ x \neq 0.$$

This assumption ensures the same ergodicity and exponential convergence for large times as in the scalar case. Moreover we get existence of a unique invariant probability measure which is Gaussian with covariance Σ given by

$$\Sigma = \int_0^\infty e^{sA}BB^*e^{sA^*}\,ds.$$

Notice that the fact that Σ solves the equation

$$\Sigma A^* + A\Sigma = BB^*$$

is the crucial property of the matrix Σ. Starting from the invariant measure μ, the process $\{X_t\}_{t\geq 0}$ becomes a stationary mean-zero (Markov) Gaussian process. Its distribution is determined by its covariance function:

$$\Gamma(s,t) = \mathbb{E}_\mu\{X_s \otimes X_t\} = \begin{cases} \Sigma e^{(t-s)A^*} & \text{whenever } 0 \leq s \leq t \\ e^{(s-t)A}\Sigma & \text{whenever } 0 \leq t \leq s. \end{cases} \qquad (4.30)$$

One sees that the formulae can become cumbersome because of the possibility that A is not self-adjoint and the matrices A and Σ do not commute. However, in many of the situations which have been considered, the matrix A is self-adjoint and commutes with BB^*. In this case, we can rewrite the above formulae in an easier form. For example, the formula for Σ becomes:

$$\Sigma = BB^* \int_0^\infty e^{2sA}\,ds = BB^*(-2A)^{-1}$$

and the equality (4.30) between matrices can be rewritten in terms of the entries of the matrices. More precisely, if $x, y \in \mathbb{R}^n$ we have:

$$\langle x, \Gamma(s,t)y\rangle = \mathbb{E}_\mu\{\langle x, X_s\rangle\langle y, X_t\rangle\} = \langle\sqrt{\Sigma}x, \sqrt{\Sigma}e^{|t-s|A}y\rangle = \langle x, e^{|t-s|A}y\rangle_\Sigma$$

if we use the notation:

$$\langle\,\cdot\,,\,\cdot\,\rangle_D = \langle\sqrt{D}\,\cdot\,, \sqrt{D}\,\cdot\,\rangle = \langle D\,\cdot\,,\,\cdot\,\rangle$$

for the inner product in $\mathbb{R}^n$ naturally associated with a positive definite matrix D. The Markov process $\{X_t\}_{t\geq 0}$ is symmetric in the sense of the theory of symmetric Markov processes developed by Fukushima and his followers. Indeed, a simple integration by parts shows that it is the process associated with the Dirichlet form of the measure μ, i.e. to the quadratic form:

$$Q(f,g) = \int_{\mathbb{R}^n} \langle \nabla f(x), \overline{\nabla g(x)} \rangle_{B^*B}\, \mu_\Sigma(dx)$$

defined on the subspace $\mathcal{Q}$ of $L^2(\mathbb{R}^n, \mu_\Sigma(dx))$ comprising the absolutely continuous functions whose first derivatives (in the sense of distributions) are still in the space $L^2(\mathbb{R}^n, \mu_\Sigma(dx))$.

If we let $\{e_i\}_i$ be the orthonormal basis of $\mathbb{R}^n$ composed of the common set of eigenvectors of the commuting self-adjoint matrices A and BB^*, then the coordinate processes $X_t^{(i)} = \langle e_i, X_t\rangle$ are scalar OU processes satisfying the SDE

$$dX_t^{(i)} = -a_i X_t^{(i)} dt + b_i dw_t^{(i)}$$

where $-a_i$ and b_i^2 are the i-th eigenvalues of A and BB^* respectively and the $\{w_t^{(i)}\}_{t\geq 0}$ are independent scalar Wiener processes. In this case, the covariance Σ of the invariant measure is diagonal with respect to the basis $\{e_i\}_i$ with i-th diagonal entry given by $b_i^2/2a_i$. An infinite dimensional analog of the above situation arises in Chap. 7 in the analysis of a "random string" term structure rate model.

Even if we drop the assumptions that A is self-adjoint and commutes with BB^*, we have seen above that the existence and uniqueness of an invariant measure is guaranteed as long as the drift matrix A is strictly dissipative. However, if there exists an $x \in \mathbb{R}^n$ such that $\langle Ax, x\rangle \geq 0$ then an invariant measure may fail to exist, and even if an invariant measure exists, it may not be unique.

An important case is when A is singular, but satisfies $\langle Ax, x\rangle < 0$ for all vectors x not in the kernel of A. Notice that under this assumption, A is still dissipative though not strictly dissipative any longer, and this implies that $\ker(A) = \ker(A^*)$. Now consider the SDE:

$$dX_t = AX_t dt + BdW_t.$$

In this case, the dynamics decouple. Let $X_t = X_t^{(K)} + X_t^{(R)}$ and $B = B^{(K)} + B^{(R)}$ where $X^{(K)}$ and $X^{(R)}$ and $B^{(K)}$ and $B^{(K)}$ are the projections of X_t and B onto the kernel and range of A respectively. Then the SDE becomes the system:

$$\begin{aligned} dX_t^{(K)} &= B^{(K)} dW_t \\ dX_t^{(R)} &= AX_t^{(R)} dt + B^{(R)} dW_t. \end{aligned}$$

Clearly, there exists an invariant measure if and only if the range of B is contained in the range of A. If there exists an invariant measure, then $X_t^{(K)} = X_0^{(K)}$ for all $t \geq 0$ and hence there are an infinite number of invariant measures. These measures are convolutions of the form $\mu_\infty = \nu * N(0, \Sigma)$. The measure ν is the distribution of $X_0^{(K)}$ and can be an arbitrary measure supported on the kernel of A. The measure $N(0, \Sigma)$ is a Gaussian measure on $\mathbb{R}^n$ supported on the range of A, where the covariance is $\Sigma = \int_0^\infty e^{At} BB^* e^{A*t} dt$ as before. Most of these measures will not be Gaussian since the distribution of $X_0^{(K)}$ can be arbitrary. We will come across the same phenomenon in Chap. 7 in the analysis of the Gauss-Markov HJM models.

4.5.2 Infinite Dimensional OU Processes

We now consider the problem of the definition of OU processes in an infinite dimensional setup. The three approaches discussed in the finite dimensional case are possible. We consider them in the three bullet points below. But first, we list some of the adjustments which are needed to get to a genuine infinite dimensional setting.

1. The Euclidean space $\mathbb{R}^n$ is replaced by a (possibly infinite dimensional) Hilbert space, say F.
2. The $n \times d$ dispersion matrix B is replaced by a (possibly unbounded) operator which we will still denote by B.
3. The $n \times n$ drift coefficient matrix A is replaced by a (possibly unbounded) operator on F which we will still denote by A.
4. The role played by the d-dimensional Wiener process W can be played by an infinite dimensional Wiener process (possibly cylindrical) in the domain of the dispersion operator B. In fact, because this situation can be quite singular, the operator B and the Wiener process are sometimes bundled together, and the role of BdW_t can be played by an F-valued white noise measure with covariance given by the operator B^*B. If that is the case, the appropriate mathematical object is a linear function, say W_B, from the tensor product $L^2([0,\infty), dt) \otimes F_B$ into a Gaussian subspace of $L^2(\Omega, \mathcal{F}, \mathbb{P})$ where $(\Omega, \mathcal{F}, \mathbb{P})$ is the complete probability space we work with. If $x \in F_B$ and $t \geq 0$, then $W_B(\mathbf{1}_{[0,t)}(\cdot)x)$ should play the same role as $\langle x, W_t \rangle$ in the finite dimensional case. We shall refrain from using this approach in this book.

We now discuss the three approaches alluded to above.

Solution of a stochastic partial differential equation

In the same way the finite dimensional OU processes were defined as solutions to some specific stochastic differential equations, the infinite dimensional OU processes can be introduced as solutions of stochastic partial differential equations (SPDE for short). Formally we try to solve:

$$\frac{dX_t}{dt} = AX_t + \frac{dW_B}{dt} \tag{4.31}$$

which can be regarded as an SPDE when A is a partial differential operator. Notice that the dispersion operator B is now included in the infinite dimensional analog of the driving Wiener process. This equation can be given a rigorous meaning by considering the integral equation:

$$\langle f^*, X_t\rangle = \langle f^*, X_0\rangle + \int_0^t \langle A^* f^*, X_s\rangle ds + W(\mathbf{1}_{[0,t)}(\cdot)f^*)$$

where $W(\mathbf{1}_{[0,t)}(\cdot)f^*)$ is a rigorous way to introduce the anti-derivative of the white noise appearing in the right-hand side of Eq. (4.31). Such an equation makes sense when f^* belongs to the domain $\mathcal{D}(A^*)$ of the adjoint operator A^*. This integral equation can be solved under some restrictive conditions on the data (F, A, B, W_B). See for example [45], [132] or [3] and the references therein. All these works assume that the operator A is self-adjoint and negative. We shall consider in the next chapter situations in which the operator A satisfies neither of these assumptions. In other words, the operator A will be neither self-adjoint, nor bounded from above.

Symmetric Process Associated with a Dirichlet Form

We saw that when the finite dimensional matrix A is symmetric and negative definite, the Markov process $\{X_t\}_{t\geq 0}$ can be characterized as the symmetric process associated with the Dirichlet form of the invariant measure μ. Moreover, using the theory of Dirichlet forms and starting from the invariant measure μ and a specific notion of gradient operator, this procedure can be reversed and the process itself could be constructed from these data. This approach can be generalized to the infinite dimensional setting as well. We can assume that F is a real separable Hilbert space containing the RKHS H_Σ (recall that the self-adjoint operator Σ is defined as $BB^*(-2A)^{-1}$) and such that the canonical cylindrical measure of H_Σ extends into a sigma-additive probability measure on F. Notice that this countably additive extension is the measure which we denoted by μ_Σ so far. The abstract theory of Dirichlet forms gives a construction, starting from the Dirichlet form

$$Q_\mu(f, g) = \int_F \langle \nabla f(x), \overline{\nabla g(x)}\rangle_{BB^*}\ \mu_\Sigma(dx)$$

associated with the measure μ_Σ, of a symmetric strong Markov process $\{X_t\}_{t\geq 0}$. This construction is carried out when the dispersion operator B is the identity in the review article [31], but it can be adapted to apply to more general cases.

Gaussian Process

Infinite dimensional OU processes can also be constructed directly as Gaussian processes. See [3]. If we assume that A and B are self-adjoint and that they commute, the construction of the process starting from the origin, say $\{X_t^{(0)}\}_{t\geq 0}$, can be performed by first choosing a Hilbert space F and then by constructing a mean-zero F-valued Gaussian process with continuous sample paths and with covariance function given by:

$$\mathbb{E}\{f^*(X_s^{(0)})g^*(X_t^{(0)})\} = \int_0^{s\wedge t} \langle e^{(s-u)A}f^*, e^{(t-u)A}g^*\rangle_{BB^*}\, du$$

for all the elements f^* and g^* of the dual F^* of F. The process $\{X_t^{(f)})\}_{t\geq 0}$ starting from a generic point f of F is then constructed via the formula:

$$X_t^{(f)} = e^{tA}f + X_t^{(0)}.$$

It is plain to check that the measure μ_Σ introduced earlier is invariant for this process and that the desired process is the process started from this invariant measure. In other words, the stationary OU process we are looking for is given as the mean-zero (continuous) stationary Gaussian process with covariance:

$$\mathbb{E}\{f^*(X_s^{(0)})g^*(X_t^{(0)})\} = \langle e^{|t-s|A}f^*, g^*\rangle_{BB^*}.$$

The construction of infinite dimensional OU processes as Gaussian processes is not artificial. Indeed, in applications related to Kolmogorov's models of turbulent flows, these processes are given as multiparameter stationary and homogeneous Gaussian fields whose distributions are derived from their spectral characteristics. Since we saw that multiparameter random fields lead naturally to function space valued processes, it appears quite natural to construct them following the directives outlined above. The interested reader is referred to the Notes & Complements at the end of the chapter for specific references.

4.5.3 The SDE Approach in Infinite Dimensions

Since solving SDE in (possibly) infinite dimensional spaces will be the technique of choice to guarantee the existence of the stochastic models for the term structure of interest rate, we devote this last subsection to a quick review of this approach.

We are attempting to construct a stochastic process $X = \{X_t;\, t \geq 0\}$ on a stochastic basis $(\Omega, \mathcal{F}, (\mathcal{F}_t)_{t\geq 0}, \mathbb{P})$ satisfying the usual assumptions, and as explained earlier in this book, we assume that the state space of the process is a real separable Hilbert space F. The *drift operator* A is usually an *unbounded* operator on F. But since the result of the construction in the finite dimensional case could be expressed in terms of the semigroup e^{tA}, we assume that:

- A is the infinitesimal generator of a strongly continuous semigroup $\{e^{tA}; t \geq 0\}$ of bounded operators on F. The domain of A,

$$\mathcal{D}(A) = \left\{ f \in F;\ \lim_{t \to 0} \frac{e^{tA} f - f}{t} \text{ exists} \right\},$$

is generally a proper subset of F, which can be small, but which is always dense in F. See Sect. 4.6 below for the definitions of strongly continuous semigroups.

- The source(s) of noise are captured by a Wiener process $W = \{W_t;\ t \geq 0\}$ in a real separable Banach space E, and we denote by H the RKHS of the distribution of W_1. We have the following abstract Wiener space diagram:

$$\begin{array}{ccc} E^* & \hookrightarrow & H^* \\ & & \updownarrow \ \text{(Riesz identification)} \\ & & H \ \hookrightarrow E \end{array}$$

- The dispersion operator B is assumed to be a (deterministic) bounded linear operator from E into F. Consequently, its restriction to H is Hilbert–Schmidt.
  Note that it is possible to take B defined only on H if we start with a cylindrical Wiener process in H. But in this case, we would need to assume that B is a Hilbert–Schmidt operator.

Formally the Ornstein–Uhlenbeck process satisfies the stochastic differential equation

$$dX_t = AX_t dt + BdW_t \tag{4.32}$$

or in integral form

$$X_t = X_0 + \int_0^t AX_s ds + \int_0^t BdW_s. \tag{4.33}$$

Everything has been done for the second integral to make sense, but the first one is still a problem since X_s may not be (and presumably will not be) in the domain of A when A is unbounded. In particular, if we want to use the Picard iteration scheme directly to prove existence and uniqueness of a solution of (4.32), we need to set $X_t^{(0)} = X_0$ and

$$X_t^{(n+1)} = X_t^{(n)} + \int_0^t AX_s^{(n)} ds + \int_0^t BdW_s$$

but this would not work because $X^{(n)}$ will very likely be outside the domain $\mathcal{D}(A)$. There are several ways to define a solution of (4.32) or (4.33). A first possibility is to demand that a solution satisfy (4.32) in the sense of Schwartz distributions. This leads to the notion of a weak solution. In order to define the concept of a weak solution we remark that if $\{X_t\}_t$ satisfies (4.33), then for all $f^* \in \mathcal{D}(A^*)$ we have:

$$\begin{aligned}\langle f^*, X_t\rangle &= \langle f^*, X_0\rangle + \left\langle f^*, \int_0^t AX_s ds\right\rangle + \langle f^*, \int_0^t BdW_s\rangle\\ &= \langle f^*, X_0\rangle + \int_0^t \langle f^*, AX_s\rangle ds + \int_0^t \langle f^*, BdW_s\rangle\\ &= \langle f^*, X_0\rangle + \int_0^t \langle A^* f^*, X_s\rangle ds + \int_0^t \langle B^* f^*, dW_s\rangle.\end{aligned}$$

With this in mind we say that $\{X_t\}_t$ is a weak solution if for all $f^* \in \mathcal{D}(A^*)$ we have:

$$\langle f^*, X_t\rangle = \langle f^*, X_0\rangle + \int_0^t \langle A^* f^*, X_s\rangle ds + \int_0^t \langle B^* f^*, dW_s\rangle.$$

This notion of a weak solution should not to be confused with the probabilist's notion of weak solution of an SDE. Weak solutions are sometimes difficult to come by. So we consider a notion of a solution for a more regular form of the equation. The latter is obtained by applying the method of the variation of the constant (considering BdW_t as a forcing term) to the solution of the deterministic (infinite dimensional) dynamical system obtained by setting $B = 0$. This form of the equation is often called the *evolution form* of the equation. It reads:

$$X_t = e^{tA}X_0 + \int_0^t e^{(t-s)A}BdW_s.$$

Notice that, since X_t does not appear in the right-hand side (except for its initial value X_0) this form actually gives the solution (if any) of the equation. What do we need for the evolution form to make sense? The first term $e^{tA}X_0$ is okay if $\{e^{tA}\}_t$ is a strongly continuous semigroup on F and $X_0 \in F$ a.s. The problem of the existence of the second integral $\int_0^t e^{(t-s)A}BdW_s$ should be easy because the integrand is deterministic:

$$E \xrightarrow{B} F \xrightarrow{e^{(t-s)A}} F.$$

Again, everything is okay if B is really defined on E and is a bounded operator from E into F. Notice that we need an extra condition (for instance that B is Hilbert–Schmidt) if W is only cylindrical and B is only defined on H! Notice that the process $\{X_t\}_{t\geq 0}$ defined by the evolution equation is not necessarily an F-valued Itô process.

As explained in the previous section, because of the form of the Eq. (4.32) and because of the uniqueness of the solution, the solution $\{X_t\}_t$ has the Markov property. The transition probability is given by:

$$\begin{aligned}P_t(f, U) &= \mathbb{P}\{X_t \in U | X_0 = f\}\\ &= \mathbb{P}\left\{\int_0^t e^{(t-s)A}BdW_s \in U - e^{tA}f\right\}.\end{aligned}$$

This probability distribution results from the shift by $e^{tA}f$ of the distribution of the random vector $\int_0^t e^{(t-s)A}BdW_s$. Since the latter is a mean-zero Gaussian measure on F, it is completely characterized by its covariance operator. For $f^*, g^* \in F^*$ define:

$$\Gamma_t(f^*, g^*) = \mathbb{E}\left\{\left\langle f^*, \int_0^t e^{(t-s)A}BdW_s\right\rangle \left\langle g^*, \int_0^t e^{(t-s)A}BdW_s\right\rangle\right\}.$$

By polarization, this bilinear form is entirely determined by the quadratic form:

$$\begin{aligned}
\Gamma_t(f^*, f^*) &= \mathbb{E}\left\{\langle f^*, \int_0^t e^{(t-s)A}BdW_s\rangle^2\right\} \\
&= \int_0^t \|B^*(e^{(t-s)A})^* f^*\|_{H^*_W} ds \\
&= \int_0^t \langle f^*, e^{uA}BB^*(e^{(uA})^* f^*\rangle du \\
&= \left\langle f^*, \int_0^t e^{uA}BB^*(e^{(uA})^* du f^*, \right\rangle
\end{aligned}$$

where H_W is the RKHS associated with the law of W_1. This computation is identical to its finite dimension analog. The ergodic properties of the process can often be derived from the large time behavior of the process. So we address the question of the existence of $\lim_{t\to\infty} X_t$ in distribution. The process has an invariant measure μ_∞ if and only if the operator Σ defined by the improper integral:

$$\Sigma = \int_0^\infty e^{Au}BB^*e^{A^*u}\, du$$

is the covariance operator of a Gaussian measure on F, that is, if Σ is of trace class on F. Typically, an operator A is said to be of trace class if it can be written in the form $A = BB^*$ for a Hilbert–Schmidt operator B. The terminology comes from the fact that these operators are compact and if we denote by λ_n the eigenvalues, then we have $\sum_n |\lambda_n| < \infty$. In other words, the infinite series which should be giving the *trace* of A does converge, i.e. is finite. As in the finite dimensional case, a sufficient condition for the existence and uniqueness of an invariant measure is that A is bounded from above in the sense that there exists an $a > 0$ such that

$$\langle Af, f\rangle \leq -a|f|^2$$

for all $f \in \mathcal{D}(A)$. However, if the kernel of A is nontrivial, then the existence and uniqueness of an invariant measure are no longer guaranteed. This problem has been investigated in many specific models. See the Notes & Complements for references. We shall revisit it in the case of the interest models discussed in Chap. 7.

4.6 Stochastic Differential Equations

In this section we discuss a natural extension of the classical existence and uniqueness theory of solutions of stochastic differential equations (SDE for short) to the case of Banach space valued Wiener processes. References to the original results and their proofs are given in the Notes & Complements at the end of the chapter.

Let $\{S_t\}_{t\geq 0}$ be a strongly continuous semigroup of bounded operators on a real separable Hilbert space F. This means that the collection $\{S_t\}_{t\geq 0}$ of operators satisfies:

1. $S_0 = I$ where I is the identity on F,
2. $S_s \circ S_t = S_{t+s}$ for all $s, t \geq 0$,
3. $\lim_{t\to 0} \|S_t f - f\| = 0$ for all $f \in F$.

Given a strongly continuous semigroup $\{S_t\}_{t\geq 0}$, the infinitesimal generator A is defined by its domain $\mathcal{D}(A)$

$$\mathcal{D}(A) = \left\{ f \in F; \lim_{t\to 0} \frac{S_t f - f}{t} \text{ exists} \right\}$$

which appears as a dense subspace of F because of assumption 3 above, and its action on $\mathcal{D}(A)$ given by

$$Af = \lim_{t\to 0} \frac{S_t f - f}{t} \quad \text{for} \quad f \in \mathcal{D}(A).$$

For the sake of convenience we state the Hille–Yosida Theorem: An operator A is the infinitesimal generator of a strongly continuous semigroup $\{S_t\}_{t\geq 0}$ on F with $\|S_t\|_{\mathcal{L}(F)} \leq Me^{ct}$ if and only if the domain $\mathcal{D}(A)$ is dense, the resolvent operator $(\lambda I - A)^{-1}$ is bounded for $\lambda > c$ and the estimate $\|(\lambda I - A)^{-k}\| \leq M(\lambda - c)^{-k}$ holds for all natural numbers k.

We are interested in the solutions, if any, of the stochastic differential equation

$$dX_t = (AX_t + a(t, X_t))dt + b(t, X_t)dW_t \tag{4.34}$$

as processes $\{X_t\}_{t\geq 0}$ taking values in F. The drift is separated into two pieces: a linear but possibly unbounded operator A and a smooth but possibly nonlinear function $a : \mathbb{R}_+ \times F \hookrightarrow F$. The noise is modeled by a Wiener process $\{W_t\}_{t\geq 0}$ which we will assume is defined cylindrically on a real separable Hilbert space G. Since the Wiener process is only cylindrical, we assume that the function $b : \mathbb{R}_+ \times F \to \mathcal{L}_{\text{HS}}(G, F)$ takes values in the space of Hilbert–Schmidt operators from G into F.

If A is a bounded operator then $\mathcal{D}(A) = F$ and the natural notion of solution to Eq. (4.34) is that of the strong solution:

$$X_t = X_0 + \int_0^t (AX_s + a(s, X_s))ds + \int_0^t b(s, X_s)dW_s.$$

However, as we have seen in our discussion of the infinite dimensional Ornstein–Uhlenbeck process, if the operator A is unbounded, it may be impossible to give meaning to the above integral equation.

As before, we recast the differential equation (4.34) as an integral equation by formally using the variations of the constant formula:

$$X_t = S_t X_t + \int_0^t S_{t-s} a(s, X_s) ds + \int_0^t S_{t-s} b(s, X_s) dW_s \tag{4.35}$$

where $S_t = e^{At}$. A solution of the integral equation (4.35) is a called a *mild solution* of the evolution equation (4.34). Notice that if $\{X_t\}_{t \geq 0}$ is a mild solution, implicitly the following inequality must hold

$$\int_0^t \|S_{t-s} b(s, X_s)\|^2_{\mathcal{L}_{\mathrm{HS}}(G,F)} ds < +\infty$$

almost surely in order that the stochastic integral on the right-hand side of Eq. (4.35) is meaningful. Similarly, we demand that a mild solution $\{X_t\}_{t \geq 0}$ satisfy

$$\int_0^t \|S_{t-s} a(s, X_s)\|_F ds < +\infty$$

almost surely to ensure that the Bochner integral the right-hand side of Eq. (4.35) is well defined.

One advantage to working with mild solutions rather than strong solutions is that there are cases when a mild solution exists when a strong solution does not exist. This is generally the case with the interest rate models described in Chap. 6. However, a word of caution is in order:

> *A mild solution is in general not an Itô process! Therefore, care must be taken when using the Itô calculus in infinite dimensions. In particular, Itô's formula generally does not hold for functions of mild solutions of infinite dimensional SDE.*

We now present a basic existence and uniqueness result:

Theorem 4.3. *Consider the stochastic evolution equation*

$$dX_t = (AX_t + \alpha(t, X_t))dt + \beta(t, X_t) dW_t.$$

If the operator A generates a strongly continuous semigroup $\{S_t\}_{t \geq 0}$ and if the functions $\alpha : \mathbb{R}_+ \times F \to F$ and $\beta : \mathbb{R}_+ \times F \to \mathcal{L}_{\mathrm{HS}}(G, F)$ satisfy the Lipschitz bound

$$\|\alpha(t, x) - \alpha(t, y)\|_F + \|\beta(t, x) - \beta(t, y)\|_{\mathcal{L}_{\mathrm{HS}}(G,F)} \leq K\|x - y\|_F$$

then there exists a unique, up to indistinguishability, F-valued mild solution $\{X_t\}_{t \geq 0}$ to the equation such that for all $T \geq 0$ and $p > 2$ we have

$$\mathbb{E}\{\sup_{t \in [0,T]} \|X_t\|^p\} < C_p(1 + \|X_0\|^p).$$

Proof. The existence is proven via the Picard iteration scheme. Fix $T > 0$ and let $X_t^0 = S_t X_0$ for all $t \in [0,T]$ and

$$X_t^{(n+1)} = S_t X_0 + \int_0^t S_{t-s}\alpha(s, X_s^{(n)})ds + \int_0^t S_{t-s}\beta(s, X_s^{(n)})dW_s$$

for $n \geq 0$. Fix an exponent $p > 2$. We have the inequality

$$\begin{aligned}
&\mathbb{E}\{\sup_{t\in[0,T]} \|X_t^{(n+1)} - X_t^{(n)}\|_H^p\} \\
&\quad \leq p\mathbb{E}\left\{\left\|\sup_{t\in[0,T]} \int_0^T S_{t-s}\left(\alpha(s, X_s^{(n)}) - \alpha(s, X_s^{(n-1)})\right) ds\right\|^p\right\} \\
&\quad + p\mathbb{E}\left\{\left\|\sup_{t\in[0,T]} \int_0^T S_{t-s}\left(\beta(s, X_s^{(n)}) - \beta(s, X_s^{(n-1)})\right) dW_s\right\|^p\right\}.
\end{aligned}$$

The first term of the right-hand side is easily dealt with:

$$\begin{aligned}
&\mathbb{E}\left\{\left\|\sup_{t\in[0,T]} \int_0^T S_{t-s}\left(\alpha(s, X_s^{(n)}) - \alpha(s, X_s^{(n-1)})\right) ds\right\|^p\right\} \\
&\qquad \leq C\mathbb{E}\{\| \sup_{t\in[0,T]} \|X_t^{(n)} - X_t^{(n-1)}\|^p\}
\end{aligned}$$

for a constant C. The second term is nearly as easy – just apply Proposition 4.1 to stochastic convolution:

$$\begin{aligned}
&\mathbb{E}\left\{\left\|\sup_{t\in[0,T]} \int_0^T S_{t-s}\left(\beta(s, X_s^{(n)}) - \beta(s, X_s^{(n-1)})\right) dW_s\right\|^p\right\} \\
&\qquad \leq C\mathbb{E}\{ \sup_{t\in[0,T]} \|X_t^{(n)} - X_t^{(n-1)}\|^p\}
\end{aligned}$$

for another constant C. Combining the estimates we have

$$\mathbb{E}\{\sup_{t\in[0,T]} \|X_t^{(n+1)} - X_t^{(n)}\|_H^p\} \leq C\mathbb{E}\{\sup_{t\in[0,T]} \|X_t^{(n)} - X_t^{(n-1)}\|_H^p\}$$

for a constant C which depends on the horizon $T > 0$. In particular $C(T) \to 0$ as $T \to 0$, so that we may find a horizon T_1 such that $C(T_1) < 1$. By induction we have

$$\mathbb{E}\{\sup_{t\in[0,T]} \|X_t^{(n)} - X_t^{(n-1)}\|^p\} \leq C(T_1)^n K$$

so that

$$\sum_{n=1}^{\infty} \mathbb{E}\{\sup_{t\in[0,T]} \|X_t^{(n)} - X_t^{(n-1)}\|_H^p\}^{1/p} < \infty$$

implying that the sequence of processes $\{X^n\}_n$ converges to a continuous F-valued adapted process on $[0, T_1]$. The restriction on the length of the horizon can be removed by considering the equation on the intervals $[T_1, 2T_1]$, $[2T_1, 3T_1]$, etc. □

We can easily extend this result to the case when the functions α and β are random, assuming that a predictability condition is satisfied.

Notes & Complements

The Brownian sheet appeared in a work of Jack Kiefer [84] in a functional limit theorem proven to derive asymptotic statistical test procedures. Its distribution was independently studied in J. Yeh's Ph.D. [135], and some do call the two parameter Wiener measure the Yeh measure. The work of Kennedy cited in the first section can be found in [88] and [89]. Stochastic integration in an abstract Wiener space was developed in Kuo's Ph.D. See [94] and references therein for details. His original work included an unnecessary assumption on the existence of a series of finite dimensional projections converging toward the identity. This assumption is not needed for one can use the uniform convergence in the expansion (3.3) instead. Similar (and essentially equivalent) results were obtained by Yor in [136] in the Hilbert space setting. See also the work of Gaveau [68] in this respect.

Viewing multiparameter stochastic processes as processes with a single parameter and values in a functional space is a standard procedure used in statistics (see for example the work of Kiefer [84]), mathematical physics (see for example the work of Gross and Carmona such as [28] and the references therein) and in many other domains for the pure sake of convenience (see for example [71], [3]).

The attentive reader presumably noticed the fact that most of the examples of Gaussian random fields were given by specifying a covariance function of a *tensor product* nature. The special properties of these processes/fields were analyzed by Chevet and Carmona in a series of papers on the tensor products of abstract Wiener spaces [38], [29], [34].

The anomalous measure theoretic properties of infinite dimensional Wiener processes were first noticed by Gross in [75] and by Carmona in [30].

The reader interested in measurability issues related to random operators is referred to the book [35] by Carmona and Lacroix or to Skorohod's book [125] for complements. The extension of the classical existence and uniqueness theory for SDEs with Lipschitz coefficients to the setting of an abstract Wiener space was first given in Kuo's Ph.D. See his lecture notes [94] for details, or the book of Da Prato and Zabczyk for a more recent account of the Hilbert space valued case.

To the best of our knowledge, an infinite-dimensional version of the Ornstein–Uhlenbeck process was first introduced in Ann Piech's Ph.D. in 1970. See [112]. This infinite dimensional process did not attract much attention: it appeared at that time as a mere curiosity. Things changed dramatically with the wave of interest created by the coming of age of the Malliavin calculus. Indeed the infinite dimensional Ornstein–Uhlenbeck process constructed on the classical Wiener measure was built up to play a crucial role. A series of papers devoted to its rediscovery appear as a consequence of this renewal of interest. See for example the articles of D. Stroock [127] and P.A. Meyer [104], or references found in Nualart's book [109].

Infinite dimensional Ornstein–Uhlenbeck processes also appeared as solutions of linear stochastic partial differential equations. In this respect, the first instance of such a limit theorem is presumably due to Dawson and Salehi [45] and [46] who showed that infinite dimensional Ornstein–Uhlenbeck processes appeared naturally as a limit evolution equation. They also appeared as limits of infinite particle systems in the work of Holley and Stroock [81]. Parabolic equations driven by multi-parameter white noise led to similar linear stochastic PDEs in mathematical physics, constructive quantum field theory to be specific. See for example [77] or [28]. Another incarnation of the same process also occurred in mathematical models for neurobiology [132] by J. Walsh who used it as a model for neuronal activity. See also [3]. Finally, we quote Kolmogorov's kinematic approach to fully developed turbulence as an another instance. Indeed, this theory suggests to model the velocity field of a turbulent incompressible fluid as a homogeneous Gaussian random field, and the spectra proposed on physical grounds to describe the statistics of these fields (see for example the Avellaneda and Majda proposal [4]) impose indirectly the structure of an infinite dimensional Ornstein–Uhlenbeck process. See [31] and [33] for details. For a thorough treatment of the existence and uniqueness of invariant measures for the OU process see the book of Da Prato and Zabczyk [113].

5

The Malliavin Calculus

In this chapter, we briefly describe the differential calculus on Wiener space known as the Malliavin calculus. For the purpose of this book, the main application of this calculus is the Clark–Ocone formula, which provides a rather explicit expression for the martingale representation of certain random variables. Of course, the martingale representation of a random variable is intimately related to the notion of a replicating strategy for a contingent claim. We will use this connection in Chap. 6 to obtain information about hedging portfolios for interest rate contingent claims. Other applications of Malliavin calculus in numerical finance are sketched at the end of this chapter. The first is the computation by Monte-Carlo simulation of the "Greeks," the sensitivities of derivative prices to fluctuations in market parameters. The second is the computation of certain expectations, conditioned on events that occur with probability zero. The main tool in these applications is the integration-by-parts formula of Theorem 5.2.

5.1 The Malliavin Derivative

Like measurability, differentiability can be a touchy business in infinite dimensions!

5.1.1 Various Notions of Differentiability

Let E and F be separable Banach spaces, and let $f : E \to F$ be a given, possibly nonlinear, function. There are several ways to approach the notion of derivative $f'(x)$. We will briefly review the commonly occurring definitions before introducing the Malliavin derivative operator D. The first notion of a derivative of a function on a vector space is that of the Fréchet derivative. The function f is said to be Fréchet differentiable at $x \in E$ if there exists a bounded linear operator $A \in \mathcal{L}(E, F)$ such that

$$\lim_{\|h\|_E \searrow 0} \frac{1}{\|h\|_E} \|f(x+h) - f(x) - A\, h\|_F = 0.$$

In this case, we write $A = f'(x)$ and call $f'(x)$ the Fréchet derivative at x. Recall that we use the notation $\mathcal{L}(E,F)$ to denote the space of continuous linear maps from E into F. When E and F are Banach spaces, the space $\mathcal{L}(E,F)$ is implicitly assumed to be equipped with the Banach space structure given by the uniform norm.

The second and weaker notion of derivative is the Gâteaux derivative. The function f is said to be Gâteaux differentiable at $x \in E$ in the direction $y \in E$ if there exists a vector $g \in F$ such that

$$\lim_{\epsilon \searrow 0} \frac{1}{\epsilon} \|f(x+\epsilon y) - f(x) - \epsilon g\|_F = 0.$$

In this case, we write $g = D_y f(x)$ and call $D_y f(x)$ the Gâteaux derivative in the direction y. If f is Fréchet differentiable at x then f is necessarily Gâteaux differentiable in all directions $y \in F$ and $f'(x)y = D_y f(x)$. On the other hand, Gâteaux differentiability in all directions does not imply Fréchet differentiability, even when E is finite dimensional. Some form of *uniformity* is needed. For instance, let $f : \mathbb{R}^2 \to \mathbb{R}$ be defined by

$$f(x_1,x_2) = \begin{cases} \frac{x_1^3 x_2}{x_1^4 + x_2^2} & \text{if } x \neq 0, \\ 0 & \text{if } x = 0. \end{cases}$$

Note that for every $y \in \mathbb{R}^2$ we have $D_y f(0) = 0$; however, the function f is not Fréchet differentiable since if we let h approach $(0,0)$ along the path $h = (\epsilon, \epsilon^2)$ we have

$$\lim_{\epsilon \searrow 0} \frac{1}{\sqrt{\epsilon^2 + \epsilon^4}} (f(\epsilon, \epsilon^2) - f(0)) = \frac{1}{2}.$$

Nevertheless, if f is Gâteaux differentiable in all directions, the map $y \hookrightarrow D_y f(x)$ is linear and continuous, and the map $x \hookrightarrow D.f(x)$ is continuous as a map $E \to \mathcal{L}(E,F)$, then f is Fréchet differentiable.

These two notions of differentiability are too strong for many purposes. As we have come to expect, many pathologies arise when dealing with infinite dimensional spaces that are not present with finite dimensional ones. For instance, it is easy to construct a function f on $\mathbb{R}^d$ that is infinitely differentiable, takes the value $f(x_0) = 1$ at a prescribed point $x_0 \in \mathbb{R}^d$, and has compact support. On the other hand, the situation in infinite dimensions is very different. Recall that the unit ball of an infinite dimensional Banach space E is not compact for the norm topology, and thus a compact set in E necessarily has an empty interior. Now, consider a real valued function $f : E \to \mathbb{R}$ on an infinite dimensional Banach space E. If f has compact support, then f *cannot* be continuous on E, unless $f(x) = 0$ identically! Indeed, suppose $f(x_0) = 1$ for some $x_0 \in E$. If f were continuous, there would exist an open ball B_0 containing x_0 such that $f(x) > 0$ for all $x \in B_0$. But this would be a contradiction, since B_0 is not compact.

The situation is in fact much worse than that. Indeed, for many Banach spaces, the bounded continuously differentiable functions are not dense in the space of bounded uniformly continuous functions. In fact for some spaces, including the classical Wiener space $C_0[0,1]$, the only differentiable function with bounded support is the function identically equal to zero! See the Notes & Complements section at the end of the chapter for references.

For another example that is relevant to our discussion of Chap. 3, suppose the Banach space E supports a Gaussian measure μ, and let $H_\mu \subset E$ be the associated reproducing kernel Hilbert space. The Malliavin calculus concerns functions on E that are differentiable in the directions of H_μ. It turns out that a function may be differentiable in this weak sense, and yet not even be *continuous* on E!

For example, suppose that the separable Banach space E is infinite dimensional and let $\mathcal{E}$ be its Borel sigma-field. Recall that the subspace $H_\mu \in \mathcal{E}$ is a Borel subset of E of μ-measure zero. Now fix an element $h \in H_\mu^*$, but such that h is *not* an element of E^*. (Since E is infinite dimensional, such a vector h exists). Now consider the linear functional defined by $f(x) = \langle h, x\rangle$. Clearly, f is not continuous on E, and in fact $f(x)$ is not defined for all $x \in E$. On the other hand, recall that the random variable $x \hookrightarrow f(x)$ is a Gaussian random variable defined on the probability space $(E, \mathcal{E}, \mu)$ with finite variance $\|h\|^2_{H_\mu^*}$. In particular, the quantity $f(x)$ is defined and finite for μ-almost every $x \in E$. Then for almost every $x \in E$ and every $y \in H_\mu$ we can compute the directional derivative $D_y f(x) = \langle h, y\rangle_{H_\mu}$. The derivative of the discontinuous map f is given by the constant vector $Df(x) = h \in H_\mu^*$.

For a more concrete example of the above situation, let $E = C_0[0,1]$ be the space of continuous functions vanishing at zero, and let μ be the classical Wiener measure. That is, the evaluation functionals $W_t(\omega) = \omega(t)$ define a standard Wiener process $\{W_t\}_{t\in[0,1]}$ on the probability space $(E, \mathcal{E}, \mu)$. Recall that the RKHS of the classical Wiener space is the Cameron–Martin space $H_0[0,1]$ of absolutely continuous functions on $[0,1]$ which vanish at zero with square-integrable weak derivatives. Fix a square-integrable function $h : [0,1] \to \mathbb{R}$, and consider the linear functional f defined by $f(\omega) = \int_0^1 h(t)d\omega(t)$. The function f is not defined for all $\omega \in E$ since most elements of E are not functions of bounded variation. (Of course, if h happens to be smooth enough, one could define f by the integration by parts $f(\omega) = h(1)\omega(1) - \int_0^1 \omega(t)dh(t)$). Nevertheless, since f can be identified with an element of the dual space H_0^*, the function f is well-defined μ-almost everywhere: it is just the Wiener integral $f = \int_0^1 h(t)dW_t$. In the next section, we will define the Malliavin derivative of f by the formula $D_t f(\omega) = h(t)$.

We give a precise definition of the Malliavin derivative in the next section. Rather than defining the derivative in terms of Gâteaux derivatives of functions on a Banach space E in the direction of the a Hilbert subspace H as suggested above, we choose to present the theory in terms of derivatives of functions of an isonormal process on a Hilbert space H. These approaches are equivalent in the presence of a Gaussian measure μ on E since μ is the canonical cylindrical measure on the H_μ. Recall our discussion of isonormal processes and cylindrical measures in Chap. 3.

For other notions of differentiablity in infinite dimensional spaces, see the Notes & Complements.

5.1.2 The Definition of the Malliavin Derivative

The Malliavin derivative is a linear map from a space of random variables to a space of processes indexed by a Hilbert space. Being a derivative, it is not surprising that this operator is unbounded. We take the approach of defining it first on a core, proving that the resulting operator is closable, and then extending the definition to the closure of this set in the graph norm topology. In this section, we will be working with Hilbert spaces, and therefore we freely and without comment identify a Hilbert space H with its dual H^*.

Let H be a real separable Hilbert space, and let $\{W(h)\}_{h \in H}$ be the isonormal process of H. Recall that for all $h_1, \dots, h_n \in H$, the real random variables $W(h_1), \dots, W(h_n)$ are jointly normal with mean zero and covariances given by $\langle g, h \rangle_H$. The key example of an isonormal process is the Wiener integrals $\{\int_0^T h(s) dW_s; h \in L^2([0,T]; G)\}$ where $\{W_t\}_{t \in [0,T]}$ is a Wiener process defined cylindrically on a separable Hilbert space G. We will return to this example shortly, but in the meantime we do not need the extra structure (Hölder continuous sample paths, etc). associated with the Wiener process. Throughout this section we assume that the sigma-field $\mathcal{F}$ on the underlying probability space $(\Omega, \mathcal{F}, \mathbb{P})$ is given by the completion of the sigma-field $\sigma(W(h); h \in H)$ generated by the isonormal process.

We consider random variables taking values in a real separable Hilbert space F. The core we choose is the set of smooth random variables. A random variable $\xi \in L^2(\Omega; F)$ is smooth if there exist vectors $h_1, \dots, h_n \in H$ such that

$$\xi = \psi(W(h_1), \dots, W(h_n)) \tag{5.1}$$

where the function $\psi : \mathbb{R}^n \to F$ is infinitely differentiable with all derivatives polynomially bounded. We now define the Malliavin derivative operator D on this set.

Definition 5.1. *Let ξ be the smooth random variable defined by Eq. (5.1). The Malliavin derivative of ξ is defined to be*

$$D\xi = \sum_{i=1}^{n} \frac{\partial}{\partial x_i} \psi(W(h_1), \dots, W(h_n)) \otimes h_i.$$

Note that for smooth ξ we may view $D\xi$ as a random variable valued in the tensor product Hilbert space $F \otimes H$, or equivalently, the space of Hilbert-Schmidt operators $\mathcal{L}_{\mathrm{HS}}(H, F)$. Alternatively, the derivative $D\xi$ is an F-valued stochastic process $\{D_h \xi\}_{h \in H}$ indexed by H, where

$$D_h\xi = \sum_{i=1}^n \frac{\partial}{\partial x_i}\psi(W(h_1),\dots,W(h_n))\langle h_i, h\rangle.$$

Finally, if the isonormal process is given by Wiener integrals $W(h) = \int_0^T h(t)dW_t$ with respect to a cylindrical Wiener process $\{W_t\}_{t\in[0,T]}$ on a separable Hilbert space G, then the isonormal process is indexed by the Hilbert space $H = L^2([0,T];G)$, and we may think of the derivative $D\xi$ as an $F \otimes G \simeq \mathcal{L}_{\mathrm{HS}}(G,F)$-valued stochastic process $\{D_t\xi\}_{t\in[0,T]}$ indexed by the interval $[0,T]$, where

$$D_t\xi = \sum_{i=1}^n \frac{\partial}{\partial x_i}\psi(W(h_1),\dots,W(h_n)) \otimes h_i(t).$$

In this case, the Malliavin derivative $D\xi$ is an element of $L^2(\Omega; F \otimes L^2([0,T];G)) \simeq L^2([0,T]\times\Omega; F\otimes G)$; in other words, it is strictly speaking an equivalence class of functions in (t,ω) taking values in $F\otimes G$ which agree Leb $\times\,\mathbb{P}$ almost surely, where Leb is the Lebesgue measure. By Fubini's theorem we can find a representative of $D\xi$ such that for every $t\in[0,T]$ we have that $D_t\xi$ is measurable in ω and for every $\omega\in\Omega$ we have that $D\xi(\omega)$ is measurable in t. We choose this representative to define $D\xi$.

Now to extend the domain of D, we need to prove that the operator given by Definition 5.1 can be closed. To this end we prove the first and simplest version of the integration by parts formula.

Lemma 5.1. *For deterministic $h \in H$ and real-valued smooth random variable ξ we have:*

$$\mathbb{E}\left\{\langle D\xi, h\rangle_H\right\} = \mathbb{E}\left\{\xi W(h)\right\}. \tag{5.2}$$

Proof. Let $\xi = \psi(W(h_1),\dots,W(h_n))$ for a infinitely differentiable function $\psi : \mathbb{R}^n \to \mathbb{R}$. By the Gram–Schmidt orthonormalization procedure, there is no generality lost assuming that $h = h_1$ and that $h_1,\dots,h_n$ are orthonormal.

Then by integrations by parts from ordinary calculus we have

$$\begin{aligned}
\mathbb{E}\left\{\langle D\xi, h\rangle_H\right\} &= \mathbb{E}\left\{\frac{\partial}{\partial x_1}\psi(W(h_1),\dots,W(h_n))\right\} \\
&= \frac{1}{(2\pi)^{n/2}}\int_{\mathbb{R}^n}\frac{\partial\psi}{\partial x_1}(x_1,\dots,x_n)e^{-\|x\|^2/2}dx \\
&= \frac{1}{(2\pi)^{n/2}}\int_{\mathbb{R}^n} x_1\,\psi(x_1,\dots,x_n)e^{-\|x\|^2/2}dx \\
&= \mathbb{E}\{\xi W(h)\}
\end{aligned}$$

which gives the desired formula (5.2). □

Corollary 5.1. *For a vector $h \in H$ and F-valued smooth random variables ξ and η we have*

$$\mathbb{E}\left\{\langle \eta, (D\xi)h\rangle_F\right\} = \mathbb{E}\left\{\langle \xi,\eta\rangle_F W(h) - \langle \xi, (D\eta)h\rangle_F\right\}.$$

Notice that we are viewing the random variables $D\xi$ and $D\eta$ as random Hilbert–Schmidt linear operators from H into F.

Proof. By direct computation from the definition and from the product rule for classical differential calculus, we have the product rule

$$D\langle \xi, \eta \rangle_F = (D\eta)^* \xi + (D\xi)^* \eta$$

from which the result follows. □

Proposition 5.1. *The Malliavin derivative D as defined for F-valued smooth random variables is closable on $L^2(\Omega; F)$.*

Proof. Since the operator is linear, it is enough to show that if a sequence of smooth random variables $\{\xi_n\}_n$ converges to 0 in $L^2(\Omega; F)$ and if the sequence $\{D\xi_n\}_n$ converges to a limit in $L^2(\Omega; F \otimes H)$, then this limit is necessarily 0. To show this we choose an arbitrary $h \in H$ and smooth η and note that by the corollary we have

$$\begin{aligned} \mathbb{E}\left\{\langle \eta, (D\xi_n)h \rangle_F\right\} &= \mathbb{E}\left\{\langle \xi_n, \eta \rangle_F W(h) - \langle \xi_n, (D\eta)h \rangle_F\right\} \\ &\leq \mathbb{E}\left\{\|\xi_n\|_F \left(\|\eta\|_F \, |W(h)| + \|(D\eta)h\|_F\right)\right\} \\ &\leq K \left(\mathbb{E}\{\|\xi_n\|_F^2\}\right)^{1/2} \to 0 \end{aligned}$$

where the constant

$$K = 2\left(\mathbb{E}\left\{\|\eta\|_F^2 W(h)^2 + \|(D\eta)h\|_F^2\right\}\right)^{1/2}$$

is finite because Gaussian random variables have moments of all orders.

Since linear combinations of rank-one operators of the form $\eta \otimes h$ are dense in $L^2(\Omega; F \otimes H)$ we have the weak convergence of the sequence $\{D\xi_n\}_n$ to 0. But since the sequence is assumed to converge strongly, this strong limit must be 0 also. □

By closability, we can extend the definition of the derivative operator to a larger domain.

Definition 5.2. *If ξ is the $L^2(\Omega; F)$ limit of a sequence $\{\xi_n\}_{n \geq 1}$ of smooth random variables such that $\{D\xi_n\}_{n \geq 1}$ converges in $L^2(\Omega; F \otimes H)$, we define $D\xi$ as*

$$D\xi = \lim_{n \to \infty} D\xi_n.$$

We use the notation $\mathbb{H}(F)$ for the subspace of $L^2(\Omega; F)$ where the derivative can be defined by Definition 5.2. This subspace is a separable Hilbert space for the graph norm

$$\|\xi\|_{\mathbb{H}(F)}^2 = \mathbb{E}\{\|\xi\|_F^2\} + \mathbb{E}\left\{\|D\xi\|_{F \otimes H}^2\right\}.$$

The following lemma will be used to prove a form of the chain rule for Lipschitz functions in the next section.

Lemma 5.2. *Let $\xi_n \to \xi$ converge in $L^2(\Omega; F)$ and suppose that there is a constant $C > 0$ such that for all n we have*

$$\mathbb{E}\left\{\|D\xi_n\|^2_{F\otimes H}\right\} < C. \tag{5.3}$$

Then the random variable ξ is in the domain $\mathbb{H}(F)$ of the derivative.

Proof. Assumption (5.3) implies that the sequence $\{\xi_n\}$ is bounded in $\mathbb{H}(F)$. Hence, there exists a subsequence $\{\xi_{n_k}\}_k$ that converges weakly in $\mathbb{H}(F)$. But since $\xi_{n_k} \to \xi$ converges in $L^2(\Omega; F)$ we see that the weak limit of $\{\xi_{n_k}\}_k$ is ξ, implying that $\xi \in \mathbb{H}(F)$. □

5.2 The Chain Rule

For the financial applications which we have in mind, we need a form of the chain rule for Malliavin derivatives of Lipschitz, but possibly not differentiable, functions from one vector space into another. Recall that the notion of differentiability is quite touchy in infinite dimensions, whereas the Lipschitz condition is usually straightforward to check. We will find it convenient to know that the chain rule is still true for this much larger set of functions, and in particular, we will use this lemma in the next section to prove that the solution to a stochastic differential equation is Malliavin differentiable.

Proposition 5.2. *Let F and G be separable Hilbert spaces. Given a random variable $\xi \in \mathbb{H}(F)$ and a function $\kappa : F \to G$ such that*

$$\|\kappa(x) - \kappa(y)\|_G \leq C\|x - y\|_F$$

for all $x, y \in F$, then the random variable $\kappa(\xi)$ is in $\mathbb{H}(G)$ (i.e. it is Malliavin differentiable) and there exists an $\mathcal{L}(F, G)$-valued random variable Z satisfying the bound $\|Z\|_{\mathcal{L}(F,G)} \leq C$ almost surely and such that

$$D\kappa(\xi) = ZD\xi.$$

Remark 5.1. The function κ is usually not Frechét differentiable, yet there exists a random variable Z which plays the role of the derivative $\kappa'(\xi)$ in the sense of the chain rule. If κ is Frechét differentiable, then $Z = \kappa'(\xi)$ almost surely.

Proof. According to Lemma 5.2, in order to show that $\kappa(\xi) \in \mathbb{H}(G)$ we need only to find a sequence of functions $\{\kappa_n\}_n$ such that $\kappa_n(\xi) \to \kappa(\xi)$ strongly in $L^2(\Omega; F)$ and such that $\{D\kappa_n(\xi)\}_n$ is bounded in $L^2(\Omega; F \otimes H)$.

Our strategy is to find a sequence of approximating functions that are cylindrical and twice differentiable. Let $\{e^i\}_{i=1}^{\infty}$ be a basis of F and let $\{r^i\}_{i=1}^{n}$ be a basis for $\mathbb{R}^n$. Let the projection from $\mathbb{R}^n$ to F be denoted $\ell_n = \sum_{i=1}^{n} e^i \otimes r^i$ and denote its Hilbert space adjoint by $\ell_n^* = \sum_{i=1}^{n} r^i \otimes e^i$.

For every n let $j_n : \mathbb{R}^n \to \mathbb{R}$ be a real valued twice-differentiable bounded function supported on the unit ball in $\mathbb{R}^n$ and such that $\int_{\mathbb{R}^n} j_n(x)dx = 1$, and for every $\epsilon > 0$ define the approximate identity j_n^ϵ by $j_n^\epsilon(x) = \epsilon^{-n} j_n(x/\epsilon)$. Set $\epsilon = 1/n$ and choose κ_n to be defined by the Bochner integral

$$\kappa_n(x) = \int_{\mathbb{R}^n} j_n^\epsilon(y - \ell_n^* x)\kappa(\ell_n y)dy = \int_{\mathbb{R}^n} j_n^\epsilon(y)\kappa(\ell_n \ell_n^* x - \ell_n y)dy.$$

Note that κ_n is differentiable and that

$$\begin{aligned}\mathbb{E}\{\|\kappa(\xi) - \kappa_n(\xi)\|_G^2\} &\leq \mathbb{E}\left\{\left(\int_{\mathbb{R}^n} j_n^\epsilon(y)\|\kappa(\ell_n\ell_n^*\xi - \ell_n y) - \kappa(\xi)\|_G dy\right)^2\right\} \\ &\leq \mathbb{E}\left\{\left(\int_{\mathbb{R}^n} j_n^\epsilon(y)(\|(\ell_n\ell_n^*\xi - \xi\|_F + \|y\|_{\mathbb{R}^n})dy\right)^2\right\} \\ &\leq 2C^2\mathbb{E}\{\|(\ell_n\ell_n^* - I)\xi\|_F^2\} + 2C^2/n^2 \to 0\end{aligned}$$

by the dominated convergence theorem.

We have

$$D\kappa_n(\xi) = \int_{\mathbb{R}^n} \kappa(\ell_n y) \otimes (\nabla j_n^\epsilon(y - \ell_n^*\xi)D\ell_n^*\xi)dy$$

where ∇ is the gradient in $\mathbb{R}^n$, so that

$$\mathbb{E}\left\{\|D\kappa_n(\xi)\|_{G\otimes H}^2\right\} \leq C^2\mathbb{E}\left\{\|D\xi\|_{G\otimes H}^2\right\}$$

and we can apply Lemma 5.2.

Finally, we note that $\nabla\kappa_n(\xi)$ is bounded in $L^\infty(\Omega; \mathcal{L}(F, G))$. Since $L^1(\Omega; \mathcal{L}_1(G, F))$ is separable, we can extract subsequence $\{\nabla\kappa_{n_k}(\xi)\}_k$ and find a random variable Z such that we have the weak-* convergence

$$\mathbb{E}\left\{\langle\eta, \kappa_{n_k}(\xi)D\xi\rangle_{\mathcal{L}_{\mathrm{HS}}(H,G)}\right\} \to \mathbb{E}\left\{\langle\eta, ZD\xi\rangle_{G\otimes H}\right\}$$

for every $\eta \in L^2(\Omega; G \otimes H)$ and hence $D\kappa(\xi) = ZD\xi$ as claimed. □

5.3 The Skorohod Integral

The differentiation operator D is a closed linear map from its domain $\mathbb{H}(F) \subset L^2(\Omega; F)$ into the Hilbert space $L^2(\Omega; F\otimes H)$. As such, there exists an adjoint operator, denoted δ, which maps a subspace of $L^2(\Omega; F \otimes H)$ into $L^2(\Omega; F)$ according to

$$\mathbb{E}\{\langle D\xi, \eta\rangle_{F\otimes H}\} = \mathbb{E}\{\langle\xi, \delta\eta\rangle_F\}.$$

The operator δ is not bounded, and its domain is given by the set of random variables $\eta \in L^2(\Omega; F \otimes H)$ such that there exists a $C \geq 0$ with

$$|\mathbb{E}\{\langle D\xi, \eta\rangle_{F\otimes H}\}| \leq C \left(\mathbb{E}\{\|\xi\|_F^2\}\right)^{1/2}$$

for all $\xi \in \mathbb{H}(F)$. Since the operator D is a form of gradient, the adjoint operator δ should be interpreted as a divergence. This divergence operator δ is often called the Skorohod integral . The relation to stochastic integration will become apparent at the end of this section.

To get a feel for how the Skorohod integral acts on random variables, let $\eta = X \otimes h$ where $X \in \mathbb{H}(F)$ is an F-valued differentiable random vector and $h \in H$ is a deterministic vector. By the integration by parts formula (5.2) given in Lemma 5.1 we have

$$\begin{aligned} \mathbb{E}\{\langle D\xi, X \otimes h\rangle_{F\otimes H}\} &= \mathbb{E}\{\langle (D\xi)h, X\rangle_F\} \\ &= \mathbb{E}\{\langle \xi, XW(h) - (DX)h\rangle_F\} \end{aligned}$$

for smooth ξ, and we conclude that $\delta(X \otimes h) = XW(h) - (DX)h$.

Now let us rework the last example $\eta = X \otimes h$ but with the additional assumption that X is independent of the random variable $W(h)$. Then X is the $L^2(\Omega; F)$ limit of smooth random variables of the form $X_n = \psi(W(h_1), \ldots, W(h_n))$ where $h_1, \ldots, h_n$ are orthogonal to h. By the definition of the Malliavin derivative we have

$$(DX_n)h = \sum_{i=1}^{n} \frac{\partial}{\partial x_i}\psi(W(h_1), \ldots, W(h_n))\langle h_i, h\rangle = 0$$

and hence by taking limits $\delta(X \otimes h) = XW(h) - (DX)h = XW(h)$. We will make use of this little observation to relate the Skorohod integral to the Itô integral.

For the rest of this section we focus on the case where the isonormal process is given by the Wiener integrals $\{\int_0^T h(t)dW_t; h \in L^2([0,T]; G)\}$ where $\{W_t\}_{t\in[0,T]}$ is a cylindrical Wiener process on separable Hilbert space G, defined on a probability space $(\Omega, \mathcal{F}, \mathbb{P})$ for which the sigma-field $\mathcal{F}$ is generated by the Wiener process. Introduce the filtration $\{\mathcal{F}_t\}_{t\in[0,T]}$ as the augmentation of the filtration generated by the Wiener process.

Theorem 5.1. *Let α be an $\mathcal{L}_{\mathrm{HS}}(G, F)$-valued predictable process with*

$$\mathbb{E}\left\{\int_0^T \|\alpha_s\|^2_{\mathcal{L}_{\mathrm{HS}}(G,F)} ds\right\} < +\infty.$$

Then α is in the domain of the Skorohod integral δ and

$$\delta(\alpha) = \int_0^T \alpha_s dW_s$$

where the integral on the right is an Itô integral as defined in Chap. 4.

Proof. First assume that the process α is simple:

$$\alpha_t = \sum_{i=0}^{N-1} \mathbf{1}_{(t_i,t_{i+1}]}(t)\alpha_{t_i}$$

where the $\mathcal{F}_{t_i}$-measurable random operator α_{t_i} is of finite rank

$$\alpha_{t_i} = \sum_{j=1}^{M} \alpha_{t_i}^{(j)} \otimes g^{(j)}$$

for some deterministic orthonormal vectors $\{g^{(j)}\}_j$ in G. Since the F-valued random variable $\alpha_{t_i}^{(j)}$ is independent of $\int_{t_i}^{t_{i+1}} g^{(j)} dW_s$ we have by linearity that

$$\begin{aligned}\delta(\alpha) &= \sum_{i=0}^{N-1} \sum_{j=1}^{M} \alpha_{t_i}^{(j)} \int_{t_i}^{t_{i+1}} g^{(j)} dW_s \\ &= \int_0^T \alpha_s dW_s.\end{aligned}$$

The conclusion now follows by the closedness of δ and the fact that such processes α are dense in the space of predictable processes in $L^2([0,T] \times \Omega; \mathcal{L}_{\mathrm{HS}}(G,F))$ □

This theorem explains the use of the terminology Skorohod integral: Skorohod integrals of predictable processes are Itô integrals. For a general process α in the domain of δ, it is customary to use the notation

$$\delta(\alpha) = \int_0^T \alpha_s \delta W_s.$$

The Skorohod integral is the starting point for building an anticipative stochastic integration theory. The definition of the Skorohod integral as the adjoint of the Malliavin derivative gives us a way to extend our original and rather limited integration by parts formula (5.2):

Theorem 5.2. *Let $\{\beta_t\}_{t\in[0,T]}$ be a predictable process in $L^2([0,T] \times \Omega; G)$ and let $\xi \in \mathbb{H}(F)$. We have*

$$\mathbb{E}\left\{\int_0^T \langle D_t\xi, \beta_t\rangle_{F\otimes G} dt\right\} = \mathbb{E}\left\{\left\langle \xi, \int_0^T \beta_t dW_t \right\rangle_F\right\}.$$

We will make use of this integration by parts formula in the derivation of the Clark–Ocone formula, as well as in the financial applications considered at the end of the chapter.

Since in practice it is hard to do numerics with Skorohod integrals, we end this section with a little result: some Skorohod integrals can be computed as Itô integrals after all.

Proposition 5.3. *Let $\{\beta_t\}_{t\in[0,T]}$ be a predictable, square-integrable, G-valued process and let $\xi \in \mathbb{H}(\mathbb{R})$ be such that*

$$\mathbb{E}\left\{\xi^2\left(\int_0^T \beta_t dW_t\right)^2 + \left(\int_0^T D_t\xi\beta_t dt\right)^2\right\} < +\infty.$$

Then we have

$$\int_0^T \xi\beta_t\delta W_t = \xi\int_0^T \beta_t dW_t - \int_0^T D_t\xi\beta_t dt.$$

Proof. Let $\eta \in \mathbb{H}(\mathbb{R})$. By the integration by parts formula we have

$$\begin{aligned}\mathbb{E}\left\{\eta\int_0^T \xi\beta_t\delta W_t\right\} &= \mathbb{E}\left\{\int_0^T D_t\eta\xi\beta_t dt\right\}\\ &= \mathbb{E}\left\{\int_0^T (D_t(\eta\xi) - \eta D_t\xi)\beta_t dt\right\}\\ &= \mathbb{E}\left\{\eta\left(\xi\int_0^T \beta_t dW_t - \int_0^T D_t\xi\beta_t dt\right)\right\}\end{aligned}$$

from which the desired result follows. □

5.4 The Clark–Ocone Formula

In this section, we fix a probability space $(\Omega, \mathcal{F}, \mathbb{P})$ supporting a cylindrical Wiener process $\{W_t\}_{t\in[0,T]}$ on a separable Hilbert space G, where the sigma-field $\mathcal{F}$ is generated by the Wiener process and the filtration $\{\mathcal{F}_t\}_{t\in[0,T]}$ as the augmentation of the filtration generated by the Wiener process. The isonormal process we will work with are the Wiener integrals $\{\int_0^T h(t)dW_t; h \in L^2([0,T];G)\}$. Again, we identify the dual space G^* with G without comment.

In Chap. 4 we encountered the martingale representation theorem, guaranteeing the existence of an integrand such that any random variable $\xi \in L^2(\Omega; F)$ is represented as a stochastic integral with respect to the cylindrical Wiener process. If the random variable $\xi \in \mathbb{H}(F)$ is differentiable, the following result allows us to explicitly compute the integrand in the martingale representation in terms of the Malliavin derivative of ξ.

Theorem 5.3 (Clark–Ocone formula). *For every $\mathcal{F}_T$-measurable random variable $\xi \in \mathbb{H}(F)$ we have the representation*

$$\xi = \mathbb{E}\{\xi\} + \int_0^T \mathbb{E}\{D_t\xi|\mathcal{F}_t\}dW_t.$$

Proof. Since $\xi \in L^2(\Omega; F)$, by the martingale representation given in Theorem 4.1 there exists a predictable process $\{\alpha_t\}_{t\in[0,T]} \in L^2(\Omega \times [0,T]; F \otimes G)$ such that

$$\xi = \mathbb{E}\{\xi\} + \int_0^T \alpha_t dW_t.$$

Without loss of generality we assume that $\mathbb{E}\{\xi\} = 0$ and we let $\{\beta_t\}_{t\in[0,T]}$ be a predictable process in $L^2([0,T] \times \Omega, F \otimes G)$ so that by the integration by parts formula Theorem 5.2 and Itô's isometry we have

$$\begin{aligned}\mathbb{E}\left\{\int_0^T \langle D_t\xi, \beta_t\rangle_{F\otimes G} dt\right\} &= \mathbb{E}\left\{\langle \xi, \int_0^T \beta_t dW_t\rangle_F\right\} \\ &= \mathbb{E}\left\{\langle \int_0^T \alpha_t dW_t, \int_0^T \beta_t dW_t\rangle_F\right\} \\ &= \mathbb{E}\left\{\int_0^T \langle \alpha_t, \beta_t\rangle_{F\otimes G} dt\right\}\end{aligned}$$

implying

$$\mathbb{E}\left\{\int_0^T \langle \lambda_t, \beta_t\rangle_{F\otimes H} dt\right\} = 0$$

where $\lambda_t = D_t\xi - \alpha_t$. Letting $\beta_t = \mathbb{E}\{\lambda_t|\mathcal{F}_t\}$ be the optional projection process we have

$$\mathbb{E}\left\{\int_0^T \|\mathbb{E}\{\lambda_t|\mathcal{F}_t\}\|^2_{F\otimes G} dt\right\} = 0$$

implying that $\alpha_t = \mathbb{E}\{D_t\xi|\mathcal{F}_t\}$ for almost every (t,ω) as desired. □

Finally, we will find the following version of the Leibnitz rule useful:

Proposition 5.4. *Suppose the predictable continuous square-integrable process $\{\alpha_t\}_{t\in[0,T]}$ is such that for all $t \in [0,T]$ the random variable $\alpha_t \in \mathbb{H}(F \otimes G)$ is differentiable. Then*

$$D_t \int_0^T \alpha_s dW_s = \alpha_t + \int_t^T D_t\alpha_s dW_s.$$

Note that this is just an instance of the commutation relation $D\delta\alpha = \alpha + \delta D\alpha$.

5.4.1 Sobolev and Logarithmic Sobolev Inequalities

In this subsection we show how the Clark–Ocone formula can be used to give short proofs of the Sobolev and logarithmic Sobolev inequalities.

Let $\{W(h)\}_{h\in H}$ be an isonormal process on a real separable Hilbert space H and let D denote the Malliavin derivative associated with this isonormal process. As before we denote by $\mathbb{H}$ the Hilbert space of real random variables in the domain of D such that

$$\mathbb{E}\{\xi^2\} + \mathbb{E}\{\|D\xi\|_H^2\} < +\infty.$$

The proofs below make use of the following insight: the identity of the index space H is often not relevant to the computation at hand. For instance, consider the smooth random variable

$$\xi = \psi(W(h_1), \ldots, W(h_n))$$

where $h_1, \ldots, h_n$ are orthonormal. Notice that the expected value

$$\mathbb{E}\{\|D\xi\|_H^2\} = \frac{1}{(2\pi)^{n/2}} \int_{\mathbb{R}^n} \sum_{i=1}^{n} \left(\frac{\partial}{\partial x_i} \psi(x_1, \ldots, x_n) \right)^2 e^{-\|x\|^2/2}\, dx$$

is computed irrespective of the space H.

In fact, as long as H is a separable and infinite dimensional Hilbert space, the above computation shows that the space $\mathbb{H}$ of differentiable random variables associated with the isonormal process $\{W(h); h \in H\}$ is isometrically isomorphic to the space of differentiable random variable associated with the Wiener integrals $\{\int_0^T h(s)dw_s; h \in L^2([0,T])\}$ where $T > 0$ and $\{w_t\}_{t\in[0,T]}$ is a scalar Wiener process.

So without loss of generality, we suppose that the isonormal process is given by the Wiener integrals $\{\int_0^T h(s)dw_s; h \in L^2([0,T])\}$. Let $\{\mathcal{F}_t\}_{t\in[0,T]}$ be the augmentation of the filtration generated by the scalar Wiener process $\{w_t\}_{t\in[0,T]}$. The advantage of working with this choice of H is that the Clark–Ocone formula becomes available. We now present a version of the classical Sobolev inequality:

Theorem 5.4. *For every $\xi \in \mathbb{H}$ we have*

$$\mathbb{E}\{\xi^2\} \leq \mathbb{E}\{\xi\}^2 + \mathbb{E}\{\|D\xi\|_H^2\}.$$

Equality is achieved if ξ is Gaussian; i.e. $\xi = C + W(h)$ for constants $C \in \mathbb{R}$ and $h \in H$.

Proof. By assumption, the random variable ξ is $\mathcal{F}_T$-measurable. By the Clark–Ocone formula, we have the following martingale representation:

$$\xi = \mathbb{E}\{\xi\} + \int_0^T \mathbb{E}\{D_t\xi|\mathcal{F}_t\}dw_t.$$

Hence by Itô's isometry and Jensen's inequality we have

$$\begin{aligned}\mathbb{E}\{\xi^2\} &= \mathbb{E}\{\xi\}^2 + \mathbb{E}\left\{\int_0^T \mathbb{E}\{D_t\xi|\mathcal{F}_t\}^2 dt\right\} \\ &\leq \mathbb{E}\{\xi\}^2 + \mathbb{E}\left\{\int_0^T (D_t\xi)^2 dt\right\} \\ &= \mathbb{E}\{\xi\}^2 + \mathbb{E}\{\|D\xi\|_H^2\},\end{aligned}$$

and the proof is finished. □

The following logarithmic Sobolev inequality is a significant improvement of the above result. See the Notes & Complements section for references.

Theorem 5.5. *For every $\xi \in \mathbb{H}$ the following inequality holds:*

$$\mathbb{E}\{\xi^2 \log(\xi^2)\} \leq \mathbb{E}\{\xi^2\} \log\left(\mathbb{E}\{\xi^2\}\right) + 2\mathbb{E}\{\|D\xi\|_H^2\}.$$

Equality is achieved if ξ is lognormal; i.e. $\xi = C\exp(W(h))$ for a constants $C \in \mathbb{R}$ and a vector $h \in H$.

Proof. Suppose ξ^2 is bounded and from above and let $\{M_t\}_{t\geq 0}$ be the positive continuous martingale given by $M_T = \xi^2$ and $M_t = \mathbb{E}\{\xi^2|\mathcal{F}_t\}$. By the Clark–Ocone formula we have

$$M_t = \mathbb{E}\{\xi^2\} + \int_0^t \mathbb{E}\{D_s\xi^2|\mathcal{F}_s\} dw_s.$$

On the other hand, by Itô's formula we have

$$M_t \log(M_t) = M_0 \log(M_0) + \int_0^t (1 + \log(M_s)) dM_s + \frac{1}{2}\int_0^t \frac{d \ll M \gg_s}{M_s}.$$

Combining the above formulas and the Cauchy–Schwarz inequality yields

$$\begin{aligned}\mathbb{E}\{\xi^2 \log(\xi^2)\} &= \mathbb{E}\{\xi^2\}\log\left(\mathbb{E}\{\xi^2\}\right) + \frac{1}{2}\mathbb{E}\left\{\int_0^T \frac{\mathbb{E}\{D_s\xi^2|\mathcal{F}_s\}^2}{M_s} ds\right\} \\ &= \mathbb{E}\{\xi^2\}\log\left(\mathbb{E}\{\xi^2\}\right) + 2\mathbb{E}\left\{\int_0^T \frac{\mathbb{E}\{\xi D_s\xi|\mathcal{F}_s\}^2}{M_s} ds\right\} \\ &\leq \mathbb{E}\{\xi^2\}\log\left(\mathbb{E}\{\xi^2\}\right) + 2\mathbb{E}\left\{\int_0^T \frac{\mathbb{E}\{\xi^2|\mathcal{F}_s\}\mathbb{E}\{(D_s\xi)^2|\mathcal{F}_s\}}{M_s} ds\right\} \\ &= \mathbb{E}\{\xi^2\}\log\left(\mathbb{E}\{\xi^2\}\right) + 2\mathbb{E}\{\|D\xi\|^2\}.\end{aligned}$$

To conclude the proof we get rid of the assumption of boundedness of ξ^2 by a standard localization argument based on monotone cut-offs of ξ. □

5.5 Malliavin Derivatives and SDEs

We now consider the Malliavin derivative of F-valued random variables $\xi = X_T$, where $\{X_t\}_{t\geq 0}$ is the solution of a stochastic differential equation of the form:

$$dX_t = (AX_t + a(t,X_t))dt + b(t,X_t)dW_t.$$

which we considered first in Sect. 4.6. Keeping with the spirit of this chapter, we consider the case when $\{W_t\}_{t\geq 0}$ is a cylindrical Wiener process defined on a separable Hilbert space G. Let F be a separable Hilbert space, let us assume that A generates a strongly continuous semigroup $\{S_t\}_{t\geq 0}$ on F, and let $a : \mathbb{R}_+ \times F \to F$ and $b : \mathbb{R}_+ \times F \to \mathcal{L}_{\mathrm{HS}}(G,F)$ be globally Lipschitz. From Sect. 4.6 of Chap. 4, we know that there is a unique F-valued mild solution satisfying

$$X_t = S_t X_0 + \int_0^t S_{t-s} a(s,X_s))ds + \int_0^t S_{t-s} b(s,X_s)dW_s.$$

Recall that the process $\{X_t\}_{t\geq 0}$ is not necessarily an Itô process.

First, we show that the Malliavin derivative of the solution exists.

Lemma 5.3. *For all $T \geq 0$ we have that $X_T \in \mathbb{H}(F)$.*

Proof. By Lemma 5.2 we need only to find a sequence of Malliavin differentiable random elements, say $X_T^{(n)}$, which converge toward X_T in $L^2(\Omega;F)$, and such that $DX_T^{(n)}$ is uniformly bounded in $L^2([0,T]\times\Omega;\mathcal{L}_{\mathrm{HS}}(G,F))$. A natural candidate is provided by the elements of the Picard iteration scheme: $X_t^{(0)} = S_t X_0$ and

$$X_t^{(n+1)} = S_t X_0 + \int_0^t S_{t-u} a(u,X_u^{(n)})du + \int_0^t S_{t-u} b(u,X_u^{(n)})dW_u.$$

Indeed, applying the Leibnitz rule of Proposition 5.4 to the $n+1$-th step of the scheme we obtain:

$$\begin{aligned}
D_s X_t^{(n+1)} &= S_{t-s} b(s,X_s^{(n)}) + \int_s^t S_{t-u} D_s a(u,X_u^{(n)})du \\
&\quad + \int_s^t S_{t-u} D_s b(u,X_u^{(n)})dW_u \\
&= S_{t-s} b(s,X_s^{(n)}) + \int_s^t S_{t-u} \nabla a(u,X_u^{(n)}) D_s X_u^{(n)} du \\
&\quad + \int_s^t S_{t-u} \nabla b(u,X_u^{(n)}) X_u^{(n)} dW_u.
\end{aligned}$$

Letting $M = \sup_{t\in[0,T]} \|S_t\|$ and K be the Lipschitz constants for a and b we have

$$\begin{aligned}
&\mathbb{E}\{\|D_s X_t^{(n+1)}\|^2_{\mathcal{L}_{\mathrm{HS}}(G,F)}\} \\
&\le 3M^2 \left(\mathbb{E}\{\|b(s, X_s^{(n)})\|^2_{\mathcal{L}_{\mathrm{HS}}(G,F)}\} + 2K^2 \int_s^t \mathbb{E}\left\{\|D_s X_u\|_{\mathcal{L}_{\mathrm{HS}}(G,F)}\right\} du \right) \\
&\le 3M^2 e^{6M^2K^2(t-s)} \mathbb{E}\{\|b(s, X_s^{(n)})\|^2_{\mathcal{L}_{\mathrm{HS}}(G,F)}\}
\end{aligned}$$

where the last line follows from Gronwall's lemma. Since the Picard iterates satisfy the bound

$$\sup_n \sup_{t\in[0,T]} \mathbb{E}\{\|X_t^{(n-1)}\|_F^2\} < +\infty$$

by the assumption of linear growth of b we have

$$\sup_n \mathbb{E}\left\{\int_0^T \|D_t X_T^{(n)}\|^2_{\mathcal{L}_{\mathrm{HS}}(G,F)}\right\} < +\infty,$$

completing the proof. □

5.5.1 Random Operators

We now appeal to Skorohod's theory of random operators as developed in [125]. We will apply this theory to the study of Malliavin derivatives of solutions to stochastic differential equations.

Up to now, we have been dealing with random operators which take values in a space of bounded linear operators from one Hilbert space into another. However, we will soon have the need for more general types of random operators.

Definition 5.3. *A strong random operator from H_1 into H_2 is an H_2-valued stochastic process $\{Y(h) : h \in H_1\}$ indexed by H_1 that is linear in h: $Y(f+g) = Y(f) + Y(g)$ almost surely.*

Obviously, a random variable taking values in $\mathcal{L}(H_1, H_2)$ is a random operator in this sense. A typical example of a strong random operator which does not take values in the space $\mathcal{L}(H_1, H_2)$ is constructed as follows: Let $H_1 = H_2 = \ell^2$ and let $\xi_1, \xi_2, \ldots$ be a sequence of independent standard normal random variables. Consider the random matrix

$$Y = \begin{pmatrix} \xi_1 & 0 & \ldots \\ 0 & \xi_2 & 0 \\ \vdots & 0 & \ddots \end{pmatrix}.$$

The process $\{Y(h);\, h \in \ell^2\}$ defines a strong random operator on ℓ^2 since if $h = (h_1, h_2, \ldots) \in \ell^2$

$$Y(h) = (h_1\xi_1, h_2\xi_2, \ldots)$$

is well-defined for every $h \in \ell^2$ as an ℓ^2-valued Gaussian random variable; recall the Radonification of the canonical cylindrical Gaussian measure of Chap. 3. In fact, the map $h \hookrightarrow Y(h)$ is continuous from ℓ^2 into $L^2(\Omega; \ell^2)$ since

$$\mathbb{E}\{\|Y(f) - Y(g)\|^2\} = \|f - g\|^2.$$

However, since Y is diagonal and $\sup_i |\xi_i| = +\infty$ almost surely, the strong operator Y does not define an almost sure bounded operator on ℓ^2. We will have occasion to consider the adjoint of a strong random operator. It turns out that for many strong random operators, the natural candidate for the adjoint operator is not itself a strong random operator. To handle this situation, we define the notion of weak random operator.

Definition 5.4. *We say that Z is a weak random operator from H_1 into H_2 if Z is a real-valued stochastic process $Z = \{Z(f,g); f \in H_1, g \in H_2\}$ that is linear in both f and g.*

The adjoint, then, of a strong operator Y is the weak operator Y^* defined by the formula

$$Y^*(f,g) = \langle f, Y(g) \rangle.$$

As mentioned earlier, there exists a strong operator Y such that Y^* is not a strong operator. For instance, let $H_1 = H_2 = \ell^2$ and let $\xi_1, \xi_2, \ldots$ be as sequence of independent standard normal random variables as before. Consider the random matrix

$$Y = \begin{pmatrix} \xi_1 & \xi_2 & \ldots \\ 0 & 0 & \ldots \\ \vdots & \vdots & \ddots \end{pmatrix}.$$

Now, for each $h \in \ell^2$ the random vector $Y(h) = (h_1\xi_1 + h_2\xi_2 + ..., 0, 0, ...)$ is well-defined, since the series $\sum_i h_i\xi_i$ converges almost surely by Kolmogorov's test. However, the adjoint matrix

$$Y^* = \begin{pmatrix} \xi_1 & 0 & \ldots \\ \xi_2 & 0 & \ldots \\ \vdots & \vdots & \ddots \end{pmatrix}$$

does not define a strong random operator. Indeed, the formal expression

$$Y^*(a) = a_1(\xi_1, \xi_2, \ldots)$$

does not define a random vector in ℓ^2. It does, however, define an isonormal process on ℓ^2, if $|a_1| = 1$. To derive the useful formula of the next section, we need to let strong random operators be integrands of stochastic integrals. Let H_1, H_2 and G be real separable Hilbert spaces. Recall that the space $\mathcal{L}_{\mathrm{HS}}(G, H_2)$ of Hilbert–Schmidt operators from G into H_2 has the structure of a separable Hilbert space. Therefore, we can define a strong random operator $\{Y(h)\}_{h \in H_1}$ into $\mathcal{L}_{\mathrm{HS}}(G, H_2)$ as before. Now consider a strong operator valued process $\{Y_t\}_{t\geq 0}$ where Y_t is a strong random operator from H_1 into $\mathcal{L}_{\text{HS}}(G, H_2)$ for each $t \geq 0$. If $\{Y_t\}_{t\geq 0}$ is predictable and satisfies the integrability condition $\int_0^t \|Y_s(h)\|^2_{\mathcal{L}_{\text{HS}}(G,H_2)} ds$ for each $h \in H_1$ then by setting:

$$\left(\int_0^t Y_s \cdot dW_s\right)(h) = \int_0^t Y_s(h)\, dW_s$$

we define a strong random operator $\int_0^t Y_s \cdot dW_s$ from H_1 into H_2, where $\{W_t\}_{t\geq 0}$ is a Wiener process defined cylindrically on G.

5.5.2 A Useful Formula

Since we know that $X_t \in \mathbb{H}(F)$ for all $t \geq 0$ we can conclude by Proposition 5.2 that $a(t, X_t)$ and $b(t, X_t)$ are differentiable, and by Proposition 5.4 we see that $\{D_s X_t\}_{t\in[s,T]}$ satisfies the linear equation

$$\begin{aligned} D_s X_t = S_{t-s} b(s, X_s) &+ \int_s^t S_{t-u} \nabla a(u, X_u) D_s X_u du \\ &+ \int_s^t S_{t-u} \nabla b(u, X_u) D_s X_u dW_u. \end{aligned}$$

Since the equation is linear, its solution should be obtained from the solution $Y_{s,t}$ of the auxiliary equation

$$Y_{s,t} = S_{t-s} + \int_s^t S_{t-u} \nabla a(u, X_u) Y_{s,u} du + \int_s^t S_{t-u} \nabla b(u, X_u) Y_{s,u} \cdot dW_u \quad (5.4)$$

Indeed, if the above equation has a solution in some sense, we have the useful formula

$$D_s X_t = Y_{s,t} b(s, X_s) \mathbf{1}_{\{s\leq t\}}.$$

Fortunately, Eq. (5.4) does have a solution in the sense of strong random operators. The proof of this fact can be realized by the Picard iteration scheme detailed in Chap. 4 in the proof of the existence of mild solutions of SPDEs on F.

The random operator $Y_{s,t}$ should be thought of as the derivative dX_t/dX_s of the value of a solution to an SDE with respect to an intermediate value. As such, it is a random analog of the notion of propagator or Jacobian flow.

We now present an example that illustrates the need to introduce the concept of strong random operators into the present discussion. Let $F = G = \ell^2$ and consider the differential equation

$$dX_t^{(i)} = X_t^{(i)} dw_t^{(i)}$$

with $X_0 \in \ell^2$. That is, we are considering the time-homogeneous SDE with zero drift and diffusion term $b : \ell^2 \to \mathcal{L}_{\text{HS}}(\ell^2)$ given by $b(x) = \text{diag}(x)$. The solution to the above equation is given in component form as

$$X_t^{(i)} = e^{-t/2+w_t^{(i)}} X_0^{(i)}$$

or in vector form as

$$X_t = \text{diag}(\{e^{-t/2+w_t^{(i)}}\}_i) X_0.$$

For fixed $t > 0$, the derivative $Y_{0,t} = dX_t/dX_0$ should exist in some sense and equal $Y_{0,t} = \text{diag}(\{e^{-t/2+w_t^{(i)}}\}_i)$. This $Y_{0,t}$ does not define a random continuous operator on ℓ^2 since $\sup_i\{e^{-t/2+w_t^{(i)}}\} = +\infty$ almost surely. However, the process $\{Y_{0,t}(h); h \in \ell^2\}$ does define a strong random operator as defined in the previous section.

In the next section we use the abbreviation $Y_{0,t} = Y_t$. If Y_t is invertible then $Y_{s,t} = Y_t Y_s^{-1}$. The invertibility of the strong random operator Y_t is intimately related to the invertibility of the deterministic operator S_t. Indeed, consider the deterministic linear differential equation

$$dX_t = A\ X_t dt$$

with solution $X_t = S_t X_0$. The derivative is given by $Y_t = dX_t/dX_0 = S_t$, and in this case, Y_t is invertible only if S_t is.

In the next section we deal with a finite dimensional Hilbert space F, so the operator A is bounded and the operators $S_t = e^{At}$ are invertible for all $t \in \mathbb{R}$; i.e. the semigroup $\{S_t\}_{t\geq 0}$ can be extended to a group $\{S_t\}_{t\in\mathbb{R}}$. Hence Y_t is invertible and $Y_{s,t} = Y_t Y_s^{-1}$.

5.6 Applications in Numerical Finance

We discuss two applications of the Malliavin calculus in numerical finance. Both applications use the integration by parts formula to smooth out a singular term that appears in a formal computation.

5.6.1 Computation of the Delta

We now turn to one of the first applications of Malliavin calculus to mathematical finance to appear in the literature: the computation of the "Greeks" by Monte-Carlo simulation. We discuss only the simplest case here to highlight the main ideas. Fix a probability space $(\Omega, \mathcal{F}, \mathbb{P})$ supporting a scalar Wiener process $\{w_t\}_{t\geq 0}$, and let $\{\mathcal{F}_t\}_{t\geq 0}$ be the augmentation of the filtration generated by w. Let X_t^x be the solution to the stochastic differential equation

$$dX_t = a(X_t)dt + b(X_t)dw_t$$

with initial condition $X_0^x = x$, where the functions $a : \mathbb{R} \to \mathbb{R}$ and $b : \mathbb{R} \to \mathbb{R}$ are assumed to have continuous and bounded first derivatives, and b is bounded from below by a positive constant. We are concerned here with computing the derivative

$$\Delta = \frac{\partial}{\partial x}\mathbb{E}\{g(X_T^x)\}.$$

When the process $\{X_t\}_{t\geq 0}$ is a stock price, the above quantity is the sensitivity of the price of an option with payout $g(X_T)$ to the current price of the stock. According to the Black–Scholes–Merton theory, this quantity, called the "delta," is the amount of stock that an investor should own in order to replicate the payout of the option.

One complication in computing the delta is that the option price $\mathbb{E}\{g(X_T^x)\}$ is rarely available as an explicit function of x. In practice, one often approximates the option price by Monte-Carlo simulation: that is, the practitioner (somehow) generates N independent copies $X_T^{x,(i)}$ of X_T^x and computes the empirical average

$$\mathbb{E}\{g(X_T^x)\} \approx \frac{1}{N}\sum_{i=1}^{N} g(X_T^{x,(i)})$$

which converges to the true option price as $N \to \infty$. Unfortunately, the approximation error decays as $N^{-1/2}$, far too slowly to implement the naive approximation

$$\Delta \approx \frac{1}{\epsilon N}\sum_{i=1}^{N}[g(X_T^{x+\epsilon,(i)}) - g(X_T^{x,(i)})],$$

for a small parameter ϵ. The problem is that the difference $\frac{1}{N}\sum_{i=1}^{N}[g(X_T^{x+\epsilon,(i)}) -g(X_T^{x,(i)})]$ may not be small, even if ϵ is, because the Monte-Carlo error may not cancel out. This difficulty is especially pronounced if g is discontinuous, as in the case of a digital option.

Fortunately, the integration by parts formula given by Theorem 5.2 provides an explicit random variable Π, which depends on the dynamics of X, but not the payout function g, such that

$$\Delta = \mathbb{E}\{g(X_T^x)\Pi\}. \tag{5.5}$$

This expression leads to the far superior Monte-Carlo approximation

$$\Delta \approx \frac{1}{N}\sum_{i=1}^{N} g(X_T^{x,(i)})\Pi^{(i)}.$$

To derive Eq. (5.5), suppose that g is differentiable, and the derivative is bounded. Then we have

$$\Delta = \mathbb{E}\left\{g'(X_T^x)Y_T\right\} \tag{5.6}$$

where $Y_t = \partial X_t^x/\partial x$ is the first variation process, given by $Y_0 = 1$ and

$$dY_t = Y_t(a'(X_t^x)dt + b'(X_t^x)dw_t).$$

The process $\{Y_t\}_{t\in[0,T]}$ is intimately related to the Malliavin derivative $D_t X_T$. Indeed, we have the formula

$$D_t X_T = Y_T Y_t^{-1} b(X_t).$$

In particular, for all $t \in [0, T]$ we have

$$Y_T = D_t X_T b(X_t)^{-1} Y_t$$

so that by taking the average, we have

$$Y_T = \int_0^T D_t X_T b(X_t)^{-1} Y_t \alpha(t) dt$$

where α is any deterministic function such that $\int_0^T \alpha(t)dt = 1$ and $\int_0^T \alpha(t)^2 dt < +\infty$. For instance, one can use the constant function $\alpha(t) = 1/T$. Substituting this expression into Eq. (5.6) yields

$$\begin{aligned}\Delta &= \mathbb{E}\left\{\int_0^T g'(X_T^x) D_t X_T b(X_t)^{-1} Y_t \alpha(t) dt\right\} \\ &= \mathbb{E}\left\{\int_0^T D_t g(X_T^x) b(X_t)^{-1} Y_t \alpha(t) dt\right\} \\ &= \mathbb{E}\left\{g(X_T^x) \int_0^T b(X_t)^{-1} Y_t \alpha(t) dw_t\right\}\end{aligned}$$

where the integration by parts formula is used in the last step.

Fixing an averaging function $\alpha \in L^2([0, T])$ with $\int_0^T \alpha(t)dt = 1$, we have found the desired Malliavin weight

$$\Pi = \int_0^T b(X_t)^{-1} Y_t \alpha(t) dw_t.$$

The above formula is called the Bismut–Elworthy–Li formula. A similar formula also holds for vector valued diffusions $\{X_t\}_{t\geq 0}$ under an ellipticity condition on the diffusion term b. However, the formula breaks down in infinite dimensions. In particular, the random variable $b(X_t)$ generally takes values in the space of Hilbert–Schmidt operators from G into F. If F is infinite dimensional, then $b(X_t)$ is not invertible. See nevertheless the Notes & Complements at the end of the chapter for reference to an article where this difficulty was overcome.

5.6.2 Computation of Conditional Expectations

The second application of the Malliavin calculus is the computation of conditional expectations. To motivate what follows, consider the problem of computing $\mathbb{E}\{g(X_T)|X_t = x\}$ by Monte-Carlo simulation, where x is a real number and $\{X_t\}_{t\geq 0}$ is a real valued process such that the law of the random variable X_t is continuous. One may first think to simulate many realizations of the process, and discard all but those paths that are such that $X_t = x$. But since the law of X_t is continuous, there is probability one that one will be forced to discard all of the realizations of the process! What to do then? And can we find a way to compute a conditional expectation using Monte Carlo scenarios which do not satisfy the condition?

Let ξ and η be random variables defined on a probability space $(\Omega, \mathcal{F}, \mathbb{P})$, and suppose that the law of η is continuous. The intuitive idea is to interpret the conditional expectation $\mathbb{E}\{\xi|\eta = 0\}$ as

$$\mathbb{E}\{\xi|\eta = 0\} = \frac{\mathbb{E}\{\xi\delta(\eta)\}}{\mathbb{E}\{\delta(\eta)\}}$$

where $\delta(x)$ denotes the usual Dirac delta function. The above expression is formal, but can be justified by a standard limiting argument based on the fact that:

$$\mathbb{E}\{\xi|\eta = 0\} = \lim_{\epsilon \searrow 0} \frac{\mathbb{E}\{\xi \mathbf{1}_{(-\epsilon,\epsilon)}(\eta)\}}{\mathbb{E}\{\mathbf{1}_{(-\epsilon,\epsilon)}(\eta)\}}.$$

In order to use Malliavin calculus, let us assume that the sigma-field $\mathcal{F}$ is the completion of the sigma-field and $\{\mathcal{F}_t\}_{t\geq 0}$ the filtration generated by a scalar Wiener process $\{w_t\}_{t\in[0,T]}$ defined on $(\Omega, \mathcal{F}, \mathbb{P})$. Let D denote the Malliavin derivative with respect to the Wiener process, and suppose that ξ and η are in $\mathbb{H}(\mathbb{R})$.

Now, since the delta function $\delta(x) = \frac{d}{dx}\mathbf{1}_{\{x>0\}}$ is formally the derivative of the Heaviside function, we may substitute the expression:

$$\delta(\eta) = D_s \mathbf{1}_{\{\eta>0\}} (D_s\eta)^{-1}$$

for all $s \in [0, T]$ such that $D_s\eta \neq 0$.

In our typical application, there is a $t \in (0, T]$ such that η is $\mathcal{F}_t$-measurable; that is, we have $D_s\eta = 0$ for all $s \in (t, T]$. Let assume then that $D_s\eta \neq 0$ for almost all $(\omega, s) \in \Omega \times [0, t]$. As before, let α be any process such that $\int_0^t \alpha(s)ds = 1$ so that

$$\delta(\eta) = \int_0^t D_s \mathbf{1}_{\{\eta>0\}} (D_s\eta)^{-1} \alpha(s)\, ds.$$

Again, it is sufficient to let $\alpha(s) = 1/t$ for all $s \in [0, t]$. By the integration by parts formula we have then

$$\begin{aligned}\mathbb{E}\{\xi\delta(\eta)\} &= \mathbb{E}\left\{\int_0^t \xi (D_s\eta)^{-1} \alpha(s) D_s \mathbf{1}_{\{\eta>0\}} ds\right\} \\ &= \mathbb{E}\left\{\mathbf{1}_{\{\eta>0\}} \int_0^t \xi (D_s\eta)^{-1} \alpha(s)\ \delta w_s\right\}\end{aligned}$$

where the stochastic integral must be understood in the sense of Skorohod since the integrand is usually anticipative. Combining the above formulas we have

$$\mathbb{E}\{\xi|\eta=0\} = \frac{\mathbb{E}\left\{\mathbf{1}_{\{\eta>0\}} \int_0^t \xi(D_s\eta)^{-1}\alpha(s)\ \delta w_s\right\}}{\mathbb{E}\left\{\mathbf{1}_{\{\eta>0\}} \int_0^t (D_s\eta)^{-1}\alpha(s)\ \delta w_s\right\}}.$$

Now let's apply this formula to the solution to a stochastic differential equation. Let $\{X_t\}_{t\geq 0}$ solve the SDE

$$dX_t = a(X_t)dt + b(X_t)dw_t$$

with initial condition X_0. We have $D_sX_t = Y_tY_s^{-1}b(X_s)$ for $s \in [0,t]$. Letting $\xi = g(X_T)$ and $\eta = X_t - x$ in the above formulas we have

$$\mathbb{E}\{g(X_T)|X_t = x\} = \frac{\mathbb{E}\left\{\mathbf{1}_{\{X_t>x\}} \int_0^t g(X_T)b(X_s)^{-1}Y_sY_t^{-1}\alpha(s)\ \delta w_s\right\}}{\mathbb{E}\left\{\mathbf{1}_{\{X_t>x\}} \int_0^t b(X_s)^{-1}Y_sY_t^{-1}\alpha(s)\ \delta w_s\right\}}.$$

The above formula still is not satisfactory for the sake of numerical computations since the Skorohod integral is difficult to compute. But thanks to Proposition 5.3, the Skorohod integrals in the formula can be re-expressed in terms of much better Itô integrals:

$$\begin{aligned}\int_0^t g(X_T)b(X_s)^{-1}Y_sY_t^{-1}\alpha(s)\ \delta w_s &= g(X_T)Y_t^{-1}\int_0^t Y_sb(X_s)^{-1}dw_s \\ &\quad - \int_0^t D_s(g(X_T)Y_t^{-1})Y_sb(X_s)^{-1}ds.\end{aligned}$$

Since $D_s(g(X_T)Y_t^{-1}) = g'(X_T)Y_TY_s^{-1}b(X_s)Y_t^{-1} - g(X_T)D_sY_t^{-1}$ the final ingredient is given by the formula

$$D_sY_t^{-1} = b(X_s)Y_s^{-2}Y_t^{-2}(Z_sY_t - Z_tY_s) - Y_t^{-1}b'(X_s)$$

where $\{Z_t\}_{t\geq 0}$ is the second variation process given by the equation

$$dZ_t = (b'(X_t)Z_t + b''(X_t)Y_t^2)dW_t + (a'(X_t)Z_t + a''(X_t)Y_t^2)dt$$

and $Z_0 = 0$. Just as we think of the first variation $Y_t = dX_t/dX_0$ as the derivative of the solution of the SDE with respect to its initial condition, we think of the second variation of $Z_t = d^2X_t/dX_0^2$ as the second derivative.

Notes & Complements

Most of the important results on differentiability of functions on Banach spaces were derived in the early second half of the previous century. Some of these results are not well known and the interested reader may have to go back to the original articles to find them. We strongly believe that the fundamental result was given as early as 1954 by Kurzweil in [95]. The ensuing papers by Bonic and Frampton [19] and Whittfield [134] can be consulted for complements. In keeping with the spirit of this monograph, we have considered here only differential calculus on Banach spaces. As shown in the original articles quoted above, infinite dimensions can prevent the straightforward extension of classical results of differential calculus. See for example [72] for Goodman's discussion of the existence of partitions of unity in a reasonable form. In many cases, it is necessary to use a more general calculus. For example, a locally convex vector space E can be a convenient set-up if the function $c : \mathbb{R} \to E$ is smooth whenever $e^* \circ c : \mathbb{R} \to \mathbb{R}$ is smooth for every $e^* \in E^*$. If E and F are convenient vector spaces, then a function $f : E \to F$ is called smooth if $f \circ c : \mathbb{R} \to F$ is smooth for every smooth $c : \mathbb{R} \to E$. That is, f is smooth if it maps smooth curves into smooth curves. This theory of convenient calculus can be found the book of Kriegl and Michor [92]. Filipović and Teichmann [64], [63], and [65] exploited the generality of this calculus to study in depth the finite dimensional realizations of HJM interest rate models. We shall review some results from this theory in Chap. 6.

Much of the material on Malliavin calculus in this chapter can be found in the book [109] of Nualart. A more probabilist, and less formal, introduction can also be found in the second volume of Rogers and Williams' text [118].

The notion of the derivative of functions on an abstract Wiener space in the direction of the reproducing kernel Hilbert space seems to have been introduced by Gross. It should be mentioned here that the original application of the Malliavin calculus is Malliavin's probabilistic proof of Hörmander's theorem on the regularity of the solutions of hypoelliptic PDE.

The Clark–Ocone formula seems to have first appeared in the paper [39] of Clark. In this paper, the formula is proven for random variables ξ, which are Fréchet differentiable functionals on the Banach space $E = C_0([0,T])$. Ocone [110] proved the formula under the much weaker assumption of Malliavin differentiability. An early financial application of the Clark–Ocone formula appeared in the paper of Karatzas and Ocone [111]. The version of the Clark–Ocone formula appearing in this chapter requires square-integrability. The integrabilty assumption can be significantly weakened. The reader interested in this generalization should consult the paper of Karatzas, Li, and Ocone [85].

The logarithmic Sobolev inequality in the form presented in this chapter was first derived by L. Gross in [76]. This inequality is equivalent to the following hypercontractive estimate of Nelson [108]:

$$\|e^{-tN}\|_{p,q} \leq 1$$

if $e^{-t} \leq ((q+1)/(p-1))^{1/2}$, where $\|A\|_{p,q}$ denotes the operator norm of A as an operator from $L^q(\Omega)$ into $L^p(\Omega)$, and the so-called *number operator* $N = \delta \circ D$ is the self-adjoint operator on $L^2(\Omega)$ whose quadratic form is given by $\langle N\xi, \xi\rangle_{L^2(\Omega)} = \mathbb{E}\{\|D\xi\|^2\}$. We learned the simple proof of the logarithmic Sobolev inequality using the Clark–Ocone formula from Capitaine, Hsu, and Ledoux [27].

The stochastic flow $\{Y_{s,t}\}_{0\leq s\leq t}$ provides a useful connection between the solutions of SDEs and Malliavin calculus. It has also been exploited by Elliott and van der Hoek [55] to provide an alternative approach to the affine short rate models based on the forward measure rather than the PDE techniques discussed in Chap. 2.

The section on the application of the integration by parts formula to the computation of the Greeks was inspired by the original article [67] of Fournié, Lasry,

Lebuchoux, Lions, and Touzi. In our presentation, we have glossed over many issues related to the actual implementation of a Monte-Carlo scheme to compute the delta.

Most importantly, while the error of a Monte-Carlo simulation of $\mathbb{E}\{\xi\}$ always decays as $N^{-1/2}$, where N is the number of independent realizations of ξ, the error grows as the standard deviation $\mathrm{var}(\xi)^{1/2}$. In our case, there is a lot of flexibility in constructing the Malliavin weight Π: recall that we may choose any averaging function $\alpha \in L^2([0,T])$ with $\int_0^T \alpha(t)dt = 1$. A clever implementation of a Monte-Carlo scheme should use the α which minimizes the variance.

Furthermore, it can be the case that $\mathrm{var}(g'(X_T)Y_T) < \mathrm{var}(g(X_T)\Pi)$, even for the optimal weight Π! In other words, there are cases where the integration-by-parts trick only increases the variance, and makes our Monte-Carlo simulation less efficient. For instance, this is clearly the case when g is a constant function. Indeed, the integration by parts trick only pays off if g is very singular. For a general function g, therefore, one should find a decomposition $g = g_{\mathrm{smooth}} + g_{\mathrm{singular}}$ and apply the integration by parts trick only to the singular part.

Further discussion of the hedging error produced by this method of computing the Greeks, as well as numerical comparisons with other methods can be found in the work of Cvitanic, Ma, and Zhang [42] and the recent survey [91] by Kohatsu-Higa and Montero. This Monte Carlo computation of the sensitivities of the solution of a stochastic differential equation was extended to the infinite dimensional setting of the sensitivity computations for solutions of stochastic partial differential equations in a paper by Carmona and Wang [37], where the lack of invertibility of the diffusion matrix is resolved by the linearity of the equation.

In a sequel article [66], Fournié, Lasry, Lebuchoux, and Lions extended the original idea of [67] to the computation of conditional expectations. This remarkable proposal initiated a spate of publications. First the idea was formalized in a more general setting by Bouchard, Ekeland and Touzi in [20], applications to the numerical solution of backward stochastic differential equations were given in [21] by Bouchard and Touzi, and applications to the computation of American and other exotic derivatives followed. Indeed the refinement by Longstaff and Schwartz [102] to the original proposal [130] of Tsitsiklis and Van Roy which became the industry standard to compute American options on large baskets is based on the computation by Monte-Carlo techniques of a large number of conditional expectations, and the use of a Malliavin-like approach as described above is a natural candidate for this type of algorithm. The interested reader will find related applications in a paper by Gobet and Kohatsu-Higa [69] and in the recent preprints of Elie, Fermanian and Touzi [54], and of Mrad, Touzi and Zeghal [103].

Part III

Generalized Models for the Term Structure of Interest Rates

6

General Models

The previous three chapters took us away from the main thrust of this book, namely the term structure of interest rates. We now come back to the stochastic models of a bond market of riskless zero coupon bonds, and we take advantage of our short excursion in the world of infinite dimensional stochastic analysis to generalize the classical models, and especially the HJM model, to include infinitely many factors.

6.1 Existence of a Bond Market

Throughout this chapter we assume the existence of a frictionless market (in particular we ignore transaction costs) for riskless zero coupon bonds of all maturities. As before, we follow the convention in use in the financial mathematics literature and we denote by $P(t,T)$ the price at time t of a zero coupon bond with maturity date T and nominal value \$1. So we assume the existence of a filtered probability space $(\Omega, \mathcal{F}, \{\mathcal{F}_t\}_{t\geq 0}, \mathbb{P})$ and for each $T > 0$, of a non-negative adapted process $\{P(t,T);\ 0 \leq t \leq T\}$ which satisfies $P(T,T) = 1$. We shall specify the dynamics of the bond prices in an indirect way, namely through prescriptions for the instantaneous forward rates, but as explained earlier, this is quite all right.

We assume that our bond prices $P(t,T)$ are differentiable functions of the maturity date T, and following our discussion of the first chapter, we define the instantaneous forward rates as:

$$f(t,T) = -\frac{\partial \log P(t,T)}{\partial T}$$

in such a way that:

$$P(t,T) = \exp\left(-\int_t^T f(t,s)ds\right).$$

As already explained in Chap. 2 we shall often use Musiela's notation:

$$P_t(x) = P(t,t+x) \qquad \text{and} \qquad f_t(x) = f(t,t+x), \qquad t,x \geq 0.$$

6.2 The HJM Evolution Equation

One of the goals of this chapter is to analyze the HJM equation:

$$df_t(x) = \left(\frac{\partial}{\partial x} f_t(x) + \alpha_t(x)\right) dt + \sum_{i=1}^{\infty} \sigma_t^i(x) dw_t^{(i)} \tag{6.1}$$

where $\{w^{(i)}\}_i$ are independent scalar Wiener processes, and where the drift is given by the famous HJM no-arbitrage condition

$$\alpha_t(x) = \sum_{i=1}^{\infty} \sigma_t^i(x) \left(\int_0^x \sigma_t^i(u) du + \lambda_t^i\right). \tag{6.2}$$

We would like to think of the forward rate curve $x \hookrightarrow f_t(x)$ as an $\mathcal{F}_t$-measurable random vector taking values in a function space F. Once we choose an appropriate space F, we will interpret Eq. (6.1) by rewriting it as a stochastic evolution equation in F.

We would also like to think of our Wiener process as a cylindrical Wiener process defined on a real separable Hilbert space G. As in the previous chapter, we shy away from realizing W_t as a bona fide Banach space valued Wiener process in the sense of Chap. 4 because the actual nature of the state space is irrelevant, only the reproducing kernel Hilbert space plays a role in this chapter. We do not even need any special features of the space G except that it is infinite dimensional. Because the eigenvalues of a Hilbert–Schmidt operator must decay fast enough for the sum of their squares to be finite, assuming that G is infinite dimensional does not disagree with the principal component analysis of Chap. 1 used to justify the introduction of models with finitely many factors or HJM models with finite rank volatility. No generality would be lost letting $G = \ell^2$ and the reader is free to substitute ℓ^2 everywhere G appears in what follows. Of course, choosing $G = \ell^2$ is equivalent to fixing a basis for G and working with the coordinates of vectors expressed in this basis. We prefer, though, to keep our presentation free of coordinates whenever possible. Also, keeping G unspecified allows for the possibility that the Wiener process takes values in a function space. Equivalently, the infinite dimensional Wiener process may be viewed as a two parameter random field with a tensor covariant structure. In any event, we pick our favorite G once and for all, and fix it for the remainder of the chapter. To simplify the presentation, we will always identify G with its dual G^*.

6.2.1 Function Spaces for Forward Curves

Our first job as a term structure modeler is to choose the state space F for the forward rate dynamics in such a way that the mathematical analysis of Eq. (6.1) is clean. However this space should be general enough to accommodate as large a family of models as possible. Building on our early discussions of principal component analysis of Chap. 1, and the HJM prescription in Chap. 2, we now list the assumptions that we use to carry out this analysis.

Assumption 6.1. 1. The space F is a separable Hilbert space and the elements of F are continuous, real-valued functions. The domain χ of these functions is either a bounded interval $[0, x_{\max}]$ or the half-line $\mathbb{R}_+$. We also assume that for every $x \in \chi$, the evaluation functional:

$$\delta_x(f) = f(x)$$

is well-defined, and is in fact a continuous linear function on F, i.e. an element of the dual space F^*.

2. The semigroup $\{S_t\}_{t \geq 0}$ is strongly continuous in F, where the left shift operator S_t is defined by:

$$(S_t f)(x) = f(t + x). \tag{6.3}$$

The generator of $\{S_t\}_{t \geq 0}$ is the (possibly unbounded) operator A.

3. The map F_{HJM} is measurable from some non-empty subset $D \subset \mathcal{L}_{\text{HS}}(G, F)$ into F where the HJM map F_{HJM} is defined by

$$F_{\text{HJM}}(\sigma)(x) = \langle \sigma^* \delta_x, \sigma^* I_x \rangle_G$$

for each $\sigma \in \mathcal{L}_{\text{HS}}(G, F)$, where G is a given real separable Hilbert space.

Let us remark on these assumptions. The most important property that the space F should have is that elements of F should be locally integrable functions – indeed, the formula for the bond price:

$$P(t, T) = \exp\left(- \int_0^{T-t} f_t(s)\, ds \right),$$

should make sense! For instance, the classical Lebesgue spaces $L^p(\mathbb{R}_+)$ have this property. Recall, however, that space $L^p(\mathbb{R}_+)$ is in fact a space of *equivalent classes of functions.* As such, its elements are only defined almost everywhere, and they cannot be evaluated on a set of measure zero.

In our analysis, we will find it necessary to be able to evaluate a forward curve (i.e. an element of the space F) at a given time to maturity. For instance, our discussion of Chap. 1 suggests that we define the short interest rate at time t as the value $f_t(0)$. This definition may not always be meaningful unless we assume the elements of F are genuine functions, not just equivalence classes.

Fortunately, almost everyone working with the term structure of interest rates would agree that the forward curves should be smooth functions of the time to maturity. Hence our Assumption 6.1.1 is reasonable. Of course, the elements of F are locally integrable, but more is true. The definite integration functional I_x defined by

$$I_x(f) = \int_0^x f(s)ds$$

is continuous on F for each $x \in \chi$ since

$$\begin{aligned} \left| \int_0^x f(s)ds \right| &\le x \sup_{s\in[0,x]} |f(s)| \\ &\le x \sup_{s\in[0,x]} \|\delta_s\|_{F^*} \|f\|_F \end{aligned}$$

and $\sup_{s\in[0,x]} \|\delta_s\|_{F^*}$ is finite by the Banach–Steinhaus theorem.

The financial implication of Assumption 6.1.1 is that the short interest rate r_t is well-defined as:

$$r_t = f_t(0)$$

Once the short rate is defined, the money-market account is defined by:

$$B_t = \exp\left(\int_0^t r_s ds \right)$$

It is the solution of the ordinary differential equation $dB_t = r_t B_t\, dt$ which satisfies the initial condition $B_0 = 1$. It is a traded asset that pays the floating interest rate r_t continuously compounded. We shall use it as a numeraire, i.e. the unit in which the prices of all the other assets are expressed. Prices expressed in units of the numeraire are denoted with a tilde and are called discounted prices:.

$$\tilde{P}_t(x) = B_t^{-1} P_t(x) = \exp\left(- \int_0^t r_s ds - \int_0^x f_t(y)dy \right). \tag{6.4}$$

We should mention that the assumption that F has the structure of a separable Hilbert space is motivated less by financial considerations than by mathematical convenience. In particular, the stochastic integration theory developed in Chap. 4 is available for use.

The semigroup $\{S_t\}_{t\ge 0}$ defined in Assumption 6.1.2 allows us to pass from the time *of* maturity notation $f(t,T)$ to Musiela's time *to* maturity notation $f_t(x)$ where $f_t(x) = f(t, t+x)$. Note that all of the evaluation functionals $\delta_x = S_x^* \delta_0$ can be recovered by a left shift of the functional δ_0. The connection between the shift operators and the presence of enough smooth functions relies on the fact that the shift operators form a semigroup of operators whose infinitesimal generator A should be the operator of differentiation, in the sense that one should have $Af = f'$ whenever f is differentiable.

Assumption 6.1.3 is intimately related to the no-arbitrage principle. In particular, we will need the function F_{HJM} in order to define the drift term of an abstract HJM model.

Note that since the elements of F are continuous, the function $x \hookrightarrow F_{\text{HJM}}(\sigma)(x)$ is continuous for all $\sigma \in \mathcal{L}_{\text{HS}}(G,F)$. However, it is not necessarily true that $F_{\text{HJM}}(\sigma)$ is an element of F. In fact, for the spaces we shall consider, it is generally false that $F_{\mathrm{HJM}}(\sigma)$ is an element of F unless the operator σ is an element of a proper subset $D \subset \mathcal{L}_{\mathrm{HS}}(G, F)$.

Assumption 6.1.3 is usually hard to check in practice. We give a sufficient condition which will prove useful when we analyze a concrete example in Sect. 6.3.3.

Assumption 6.2. The space F satisfies Assumption 6.1.1 and 6.1.2. Furthermore, there exists a subspace $F^0 \subset F$ such that the binary operator $\star$ defined by the formula

$$(f \star g)(x) = f(x) \int_0^x g(s)ds$$

maps $F^0 \times F^0$ into F, and is such that for all $f, g \in F^0$ the following bound holds:

$$\|f \star g\|_F \leq C\|f\|_F \|g\|_F$$

for some constant $C > 0$.

Proposition 6.1. *Let the space F satisfy Assumption 6.2. Then the map F_{HJM} satisfies the local Lipschitz bound*

$$\|F_{\mathrm{HJM}}(\sigma_1) - F_{\mathrm{HJM}}(\sigma_2)\|_F \leq C\|\sigma_1 + \sigma_2\|_{\mathcal{L}_{\mathrm{HS}}(G,F)} \|\sigma_1 - \sigma_2\|_{\mathcal{L}_{\mathrm{HS}}(G,F)}$$

for all Hilbert–Schmidt operators $\sigma_1, \sigma_2 \in \mathcal{L}_{\mathrm{HS}}(G, F^0)$ with ranges contained in F^0. In particular, the map F_{HJM} is measurable from $D = \mathcal{L}_{\mathrm{HS}}(G, F^0)$ into F.

Proof. We have the simple estimate:

$$\begin{aligned}\|f \star f - g \star g\| &= \frac{1}{2}\|(f-g) \star (f+g) + (f+g) \star (f-g)\| \\ &\leq C\|f-g\|\|f+g\|.\end{aligned}$$

Notice that the HJM function F_{HJM} is then recovered by the norm convergent series

$$F_{\mathrm{HJM}}(\sigma) = \sum_{i=1}^{\infty} (\sigma g_i) \star (\sigma g_i)$$

for $\sigma \in \mathcal{L}_{\mathrm{HS}}(G, F^0)$, where $\{g_i\}_{i \in \mathbb{N}}$ is a complete orthonormal system for G. The proof is now complete since we have

$$\begin{aligned}\|F_{\mathrm{HJM}}(\sigma_1) - F_{\mathrm{HJM}}(\sigma_2)\|_F &\leq \sum_{i=1}^{\infty} \|(\sigma_1 g_i) \star (\sigma_2 g_i) - (\sigma_2 g_i) \star (\sigma_2 g_i)\|_F \\ &= C \sum_{i=1}^{\infty} \|(\sigma_1 - \sigma_2) g_i\|_F \|(\sigma_1 + \sigma_2) g_i\|_F \\ &\leq C\|\sigma_1 - \sigma_2\|_{\mathcal{L}_{\mathrm{HS}}(G,F)} \|\sigma_1 + \sigma_2\|_{\mathcal{L}_{\mathrm{HS}}(G,F)}\end{aligned}$$

by the triangle and Cauchy–Schwarz inequalities. □

In the same way that the short rate is defined as the value of the forward rate curve at the left-hand point of the time to maturity interval $[0, x_{\max}]$, the long interest rate ℓ_t is defined as the value of the forward rate curve at the right end point of the domain χ. This is possible when $\chi = [0, x_{\max}]$ is bounded, in which case:

$$\ell_t = f_t(x_{\max}),$$

but it requires a special property of the space F when the domain $\chi = \mathbb{R}_+$ is the halfline. Indeed, in order to define:

$$\ell_t = f_t(\infty)$$

we need to make sure that, for all $f \in F$, the limit;

$$f(\infty) = \lim_{x \to \infty} f(x)$$

exists. We shall not add this to the list of properties required of F because the long rate ℓ_t will not be used in the present study. Nevertheless, the implications of the existence of the above limit will be discussed in Sect. 6.3.2. The long rate will also figure prominently in the discussion in Chap. 7 of invariant measures for the HJM model.

6.3 The Abstract HJM Model

In this section, we formulate a precise definition of an HJM model in a function space F. We assume that F satisfies Assumption 6.1.

We fix a complete probability space $(\Omega, \mathcal{F}, \mathbb{P})$ with filtration $\{\mathcal{F}_t\}_{t\geq 0}$ satisfying the usual conditions and such that there exists a Wiener process W defined cylindrically on the separable Hilbert space G. Let $\mathcal{P}$ be the predictable sigma-field on $\mathbb{R}_+ \times \Omega$. We now state a definition, which we essentially take from [62], of an HJM model for the forward rate:

Definition 6.1. *An HJM model on F is a pair of functions (λ, σ) where:*

1. *λ is a measurable function from $(\mathbb{R}_+ \times \Omega \times F, \mathcal{P} \otimes \mathcal{B}_F)$ into $(G, \mathcal{B}_G)$,*
2. *σ is a measurable function from $(\mathbb{R}_+ \times \Omega \times F, \mathcal{P} \otimes \mathcal{B}_F)$ into $(D, \mathcal{B}_{\mathcal{L}_{\mathrm{HS}}(G,F)})$,*

such that there exists a non-empty set of initial conditions $f_0 \in F$ for which there exists a unique, continuous mild F-valued solution $\{f_t\}_{t\geq 0}$ of the HJM equation:

$$df_t = (A\ f_t + \alpha(t, \cdot, f_t))\, dt + \sigma(t, \cdot, f_t) dW_t \tag{6.5}$$

where

$$\alpha(t, \omega, f) = F_{\mathrm{HJM}} \circ \sigma(t, \omega, f) + \sigma(t, \omega, f)\lambda(t, \omega, f).$$

If (σ, λ) is an abstract HJM model on the space F with initial condition $f_0 \in F$, then the forward rate process $\{f_t\}_{t\geq 0}$ satisfies the integral equation

$$f_t = S_t f_0 + \int_0^t S_{t-s}\alpha(s,\cdot,f_s)ds + \int_0^t S_{t-s}\sigma(s,\cdot,f_s)dW_s. \tag{6.6}$$

We now use Proposition 6.1 to give a sufficient condition for the existence of an HJM model.

Proposition 6.2. *Suppose that the state space F satisfies Assumption 6.2, and let the closed subspace $F^0 \subset F$ be such that $\|f \star g\|_F \leq C\|f\|_F\|g\|_F$ for $f, g \in F^0$. Assume that for every $(t,\omega,f) \in \mathbb{R}_+ \times \Omega \times F$ the range of the operator $\sigma(t,\omega,f)$ is contained in the subspace F^0. If σ is bounded and if the Lipschitz bounds*

$$\|\sigma(t,\omega,f) - \sigma(t,\omega,g)\|_{\mathcal{L}_{\mathrm{HS}}(G,F)} \leq C\|f-g\|_F$$

$$\|\sigma(t,\omega,f)\lambda(t,\omega,f) - \sigma(t,\omega,g)\lambda(t,\omega,g)\|_F \leq C\|f-g\|_F$$

are satisfied for some constant $C > 0$ and all $t \geq 0$, $\omega \in \Omega$ and $f, g \in F$, then the pair (λ, σ) is an HJM model on F. Furthermore, for any initial forward curve $f_0 \in F$ there exists a unique, continuous solution to the Eq. (6.5) such that $\mathbb{E}\{\sup_{t\in[0,T]} \|f_t\|_F^p\} < +\infty$ for all finite $T \geq 0$ and $p \geq 0$.

This result is just a consequence of Theorem 4.3 of Chap. 4.

6.3.1 Drift Condition and Absence of Arbitrage

We now fix an HJM model (σ, λ) with initial condition $f_0 \in F$, and we denote by $\{f_t\}_{t\geq 0}$ the unique solution to Eq. (6.5). To simplify the notation, let $\lambda_t = \lambda(t,\omega,f_t)$ and $\sigma_t = \sigma(t,\omega,f_t)$.

Theorem 6.1. *If we have*

$$\mathbb{E}\left\{\exp\left(-\frac{1}{2}\int_0^t \|\lambda_s\|_G^2 + \int_0^t \lambda_s dW_s\right)\right\} = 1$$

and if

$$\int_0^t \mathbb{E}\left\{\int_0^t \|\sigma_s^*\delta_{s-u}\|_G^2 du\right\}^{1/2} ds < +\infty$$

for all $t \geq 0$ then the market given by the HJM model (σ, λ) admits no arbitrage.

Proof. We compute the dynamics of the discounted bond price $\tilde{P}(t,T) = B_t^{-1}P(t,T)$. We will make use of the relation $S_\alpha^* I_u = I_{u+\alpha} - I_\alpha$, which is revealed in the following calculation:

$$(S_\alpha^* I_u)g = \int_0^u (S_\alpha g)(s)ds = \int_0^u g(s+\alpha)ds$$
$$= \int_\alpha^{u+\alpha} g(s)ds = (I_{u+\alpha} - I_\alpha)g.$$

Let us compute the dynamics of the bond price:

$$\begin{aligned}
-\log P(t,T) &= I_{T-t} f_t \\
&= I_{T-t} S_t f_0 + \int_0^t I_{T-t} S_{t-s} \alpha_s ds + \int_0^t \sigma_s^* S_{t-s} I_{T-t} dW_s \\
&= I_T f_0 - I_t r_0 + \int_0^t I_{T-s} \alpha_s ds - \int_0^t I_{t-s} \alpha_s ds \\
&\quad + \int_0^t I_{T-s} \sigma_s dW_s - \int_0^t \sigma_s^* I_{t-s} dW_s
\end{aligned}$$

Now by the stochastic Fubini theorem and the assumption of the theorem we have

$$= \int_0^t \left(\int_0^t \sigma(s)^* \delta_{t-u} dW_s \right) du = \int_0^t \sigma_s^* I_{t-s} dW_s.$$

Using $\log P(0,t) = -I_t f_0$ and

$$\int_0^t r_s(0) ds = I_t f_0 + \int_0^t I_{t-s} \alpha_s ds + \int_0^t \sigma_s^* I_{t-s} dW_s.$$

we conclude that

$$\log P(t,T) = \log P(0,t) + \int_0^t (f_s(0) - I_{T-s}\alpha_s)\, ds - \int_0^t \sigma_s^* I_{T-s} dW_s.$$

Now in $\{P(t,T)\}_{t\in[0,T]}$ is in the form of an Itô process . Applying Itô's formula yields

$$\begin{aligned}
P(t,T) = P(0,T) &+ \int_0^t P(s,T) \left(f_s(0) - I_{T-s}\alpha_s + \frac{1}{2} \|\sigma_s^* I_{T-s}\|_G^2 \right) ds \\
&- \int_0^t P(s,T) \sigma_s^* I_{T-s} dW_s
\end{aligned}$$

Finally, substituting $\alpha_s = F_{\mathrm{HJM}}(\sigma_s) + \sigma_s \lambda_s$, the discounted bond prices are given by

$$\tilde{P}(t,T) = P(0,T) - \int_0^t P(s,T) I_{T-s} \sigma_s d\tilde{W}_s$$

where $\tilde{W}_t = W_t + \int_0^t \lambda_s ds$. But the Cameron–Martin–Girsanov theorem says that there exists a measure $\mathbb{Q}$, locally equivalent to $\mathbb{P}$ such that the process $\tilde{W}_t = W_t + \int_0^t \lambda_s ds$ defines a cylindrical Wiener process on G for the measure $\mathbb{Q}$. We will find that for each $T > 0$ the discounted bond prices are local martingales under the measure $\mathbb{Q}$. Hence, by the fundamental theorem, there is no arbitrage. □

The current framework may be too general for practical needs. At this level of generality, we only know that the discounted bond prices are local martingales. They are bona-fide martingales if

$$\mathbb{E}^{\mathbb{Q}}\left\{\exp\left(-\frac{1}{2}\int_0^T \|\sigma_s^* I_{T-s}\|_G^2 + \int_0^t \sigma_s^* I_{T-s} d\tilde{W}_s\right)\right\} = 1.$$

We can ensure that the discounted bond prices are martingales if we can check the well-known Novikov condition

$$\mathbb{E}^{\mathbb{Q}}\left\{\exp\left(\frac{1}{2}\int_0^T \|\sigma_s^* I_{T-s}\|_G^2\right)\right\} < +\infty.$$

Alternatively, we can ensure the discounted bond prices are martingales if the forward rates are positive almost surely, since if the rates are positive, the discounted bond prices $\tilde{P}(t,T) = \exp(-\int_0^t f_s(0)ds - \int_0^{T-t} f_t(s)ds)$ are clearly bounded by one.

6.3.2 Long Rates Never Fall

There are some differences in modeling the forward rate as a function on a bounded interval $[0, x_{\max}]$ versus the half line $\mathbb{R}_+$. In particular, when we work on the half-line and define the long rate by the limit $\ell_t = \lim_{x\to\infty} f_t(x)$, an unexpected phenomenon is found: The long rate never falls. We give an account of this result in the context of the abstract HJM models studied in this chapter.

Let F be the state space. Throughout this subsection, we grant Assumption 6.1, as well as one additional assumption:

Assumption 6.3. Every $f \in F$ is a function $f : \mathbb{R}_+ \to \mathbb{R}$ such that the limit $f(\infty) = \lim_{x\to\infty} f(x)$ exists, and the functional $\delta_\infty : f \hookrightarrow f(\infty)$ is an element of F^*.

Fix a probability space $(\Omega, \mathcal{F}, \mathbb{P})$, and let $\{f_t\}_{t\geq 0}$ be an F-valued forward rate process given by an abstract HJM model, and let $\ell_t = f_t(\infty)$ be the long rate. We prove that the long rate is almost surely increasing.

Theorem 6.2. *For $0 \leq s \leq t$, the inequality $\ell_s \leq \ell_t$ holds almost surely.*

Proof. We use the following observation: For fixed (t, ω) we have $\lim_{T\to\infty} \tilde{P}(t,T)^{1/T} = e^{-\ell_t}$, where

$$\tilde{P}(t,T) = \exp\left(-\int_0^t f_s(0)ds - \int_0^{T-t} f_t(x)dx\right)$$

are the discounted bond prices. Since we are interested in an almost sure property of the forward rate process, we may work with any measure which is equivalent to the given measure $\mathbb{P}$. In particular, from the discussion of the previous section, there exists a measure $\mathbb{Q}$ equivalent to $\mathbb{P}$ such that the discounted bond price processes $\{\tilde{P}(t,T)\}_{t\in[0,T]}$ are local martingales simultaneously for all $T > 0$. All expected values will be calculated under this measure $\mathbb{Q}$.

Let ξ be a positive and bounded random variable. By the conditional versions of Fatou's lemma and Hölder's inequality we have

$$\begin{aligned}
\mathbb{E}\{e^{-\ell_t}\xi\} &= \mathbb{E}\{\lim_{T\to\infty} \tilde{P}(t,T)^{1/T}\xi\} \\
&= \mathbb{E}\{\mathbb{E}\{\lim_{T\to\infty} \tilde{P}(t,T)^{1/T}\xi|\mathcal{F}_s\}\} \\
&\leq \mathbb{E}\{\liminf_{T\to\infty} \mathbb{E}\{\tilde{P}(t,T)^{1/T}\xi|\mathcal{F}_s\}\} \\
&\leq \mathbb{E}\{\liminf_{T\to\infty} \mathbb{E}\{\tilde{P}(t,T)|\mathcal{F}_s\}^{1/T}\mathbb{E}\{\xi^{T/(T-1)}|\mathcal{F}_s\}^{(T-1)/T}\} \\
&\leq \mathbb{E}\{\liminf_{T\to\infty} \tilde{P}(s,T)^{1/T}\mathbb{E}\{\xi^{T/(T-1)}|\mathcal{F}_s\}^{(T-1)/T}\} \\
&\leq \mathbb{E}\{e^{-\ell_s}\xi\}.
\end{aligned}$$

We have used the fact that $\{\tilde{P}(t,T)\}_{t\in[0,T]}$ is a super-martingale for $\mathbb{Q}$. Since ξ is positive but arbitrary, the result follows. □

Notice that the above proof needs very little of the structure of the abstract HJM models introduced earlier. In fact, it is easy to see that the result holds in discrete time and with models with jumps. All that is assumed is that the long rate exists.

We note that the popular short rate models, such as the Vasicek and Cox–Ingersoll–Ross models discussed in Chap. 2, produce constant long rates. For instance, for the Vasicek model the short interest rate satisfies the SDE

$$dr_t = (\alpha - \beta r_t)dt + \sigma dw_t$$

for a scalar Wiener process $\{w_t\}_{t\geq 0}$. By the theory presented in Chap. 2 the forward rates are given by

$$f_t(x) = e^{-\beta x}r_t + (1 - e^{-\beta x})\frac{\alpha}{\beta} - \frac{\sigma^2}{2\beta^2}(1 - e^{-\beta x})^2$$

where $f_t(0) = r_t$. Note that not only the long rate is well-defined, but it is explicitly given by the constant

$$\ell_t = \frac{\alpha}{\beta} - \frac{\sigma^2}{2\beta^2}$$

independent of (t,ω).

There do exist models for which the long rate is strictly increasing. Consider an HJM model driven by a scalar Wiener process $\{w_t\}_{t\geq 0}$ with a constant volatility function given by $\sigma(x) = \sigma_0(x+1)^{-1/2}$. Since $F_{\text{HJM}} \circ \sigma(x) =$

$2\sigma_0^2(1-(x+1)^{-1/2})$ and $\int_0^t (x+t-s)^{-1/2} dw_s$ converges to zero a.s., it follows that the long rate for this model is the increasing process

$$\ell_t = \ell_0 + 2\sigma_0^2 t.$$

6.3.3 A Concrete Example

Filipović [62] proposed a family of spaces $\{H_w\}_w$ as appropriate state spaces to analyze the HJM dynamics. These spaces are defined as follows:

Definition 6.2. *For a positive increasing function $w : \mathbb{R}_+ \to \mathbb{R}_+$ such that*

$$\int_0^\infty w(x)^{-1} dx < +\infty, \tag{6.7}$$

let H_w denote the space of absolutely continuous functions $f : \mathbb{R}_+ \to \mathbb{R}$ satisfying

$$\int_0^\infty f'(x)^2 w(x) dx < +\infty$$

where f' is the weak derivative of f, which we endow with the inner product

$$\langle f, g\rangle_{H_w} = f(0)g(0) + \int_0^\infty f'(x)g'(x)w(x)dx.$$

The spaces $\{H_w\}_w$ have many nice properties as shown by the following proposition:

Proposition 6.3. *If the weight function w satisfies condition (6.7), then the inner product space H_w is a separable Hilbert space, the evaluation functional δ_x and the definite integration functional I_x defined by*

$$\delta_x(f) = f(x) \text{ and } I_x(f) = \int_0^x f(s)ds$$

are continuous on H_w for all $x \geq 0$, and furthermore, the semigroup of operators on H_w defined by

$$(S_t f)(x) = f(t+x)$$

is strongly continuous.

Proof. The evaluation functionals are uniformly bounded on H_w since

$$\begin{aligned}
|\langle \delta_x, f\rangle| = |f(x)| &= |f(0) + \int_0^x f'(u)du| \\
&\leq |f(0)| + \left(\int_0^x \frac{du}{w(u)}\right)^{1/2} \left(\int_s^\infty f'(u)^2 \, w(u) \, du\right)^{1/2} \\
&\leq C\|f\|_{H_w}.
\end{aligned}$$

Their continuity follow, and the other properties of the proposition are plain. □

Assumption 6.3 is satisfied if we choose such an H_w as state space. In particular, for every $f \in H_w$, the limit $f(\infty) = \lim_{x\to\infty} f(x)$ is well-defined as $f(\infty) = f(0) + \int_0^\infty f'(x)dx$ since the improper integral converges absolutely:

$$\int_0^\infty |f'(x)|dx \leq \left(\int_0^\infty f'(x)^2 w(x)dx\right)^{1/2} \left(\int_0^\infty \frac{dx}{w(x)}\right)^{1/2}.$$

Furthermore, the functional $\delta_\infty : f \hookrightarrow f(\infty)$ is continuous on H_w.

The most important motivation for the choice of H_w as state space is that H_w is compatible with the HJM no-arbitrage condition, at least if w increases fast enough. We will show that Assumption 6.2 is satisfied.

Proposition 6.4. *Let the subspace $H_w^0 \subset H_w$ be defined as*

$$H_w^0 = \{f \in H_w;\ f(\infty) = 0\}.$$

If the weight w satisfies the growth condition $\int_0^\infty x^2 w(x)^{-1}dx < +\infty$, then the binary operator $\star$ defined by

$$(f \star g)(x) = f(x)\int_0^x g(s)ds$$

maps $H_w^0 \times H_w^0$ into H_w^0, and satisfies

$$\|f \star g\|_{H_w} \leq C\|f\|_{H_w}\|g\|_{H_w}$$

for $C = 4\int_0^\infty x^2 w(x)^{-1}dx$. In particular, the function F_{HJM} is locally Lipschitz from $\mathcal{L}_{\mathrm{HS}}(G, H_w^0)$ into H_w^0.

Proof. We have for $f, g \in H_w^0$ the following bounds:

$$\begin{aligned}
\|f \star g\|_{H_w}^2 &= \int_0^\infty \left(f'(x)\int_0^x g(s)\ ds + f(x)g(x)\right)^2 w(x)dx \\
&\leq 2\int_0^\infty \left(f'(x)^2\|I_x\|_{H_w^{0*}}^2\|g\|^2 + \|\delta_x\|_{H_w^{0*}}^4\|f\|^2\|g\|^2\right) w(x)dx \\
&\leq 2\|f\|^2\|g\|^2\left(\sup_{x\geq 0}\|I_x\|_{H_w^{0*}}^2 + \int_0^\infty w(x)\|\delta_x\|_{H_w^{0*}}^4 dx\right).
\end{aligned}$$

Now for $f \in H_w^0$ we have the estimate

$$\begin{aligned}
|\langle I_x, f\rangle| &\leq \int_{s=0}^x \int_{t=s}^\infty |f'(t)|dsdt \\
&= \int_{t=0}^\infty (t \wedge x)|f'(t)|dt \\
&\leq \left(\int_0^\infty t^2 w(t)^{-1}dt\right)^{1/2} \|f\|
\end{aligned}$$

for all $x \geq 0$, so that $\sup_{x\geq 0} \|I_x\|_{H_w^{0*}}$ is finite. Similarly, for $f \in H_w^0$ we have

$$|f(x)| \leq \int_x^\infty |f'(s)|ds$$
$$\leq \left(\int_x^\infty w(s)^{-1}ds\right)^{1/2} \|f\|$$

so that

$$\int_0^\infty w(x)\|\delta_x\|^4 dx \leq \int_0^\infty w(x)\left(\int_x^\infty w(s)^{-1}ds\right)^2 dx$$
$$= 2\int_{s=0}^\infty \int_{t=0}^s \int_{x=0}^t w(x)w(s)^{-1}w(t)^{-1}ds\ dt\ dx$$
$$\leq \int_0^\infty s^2 w(s)^{-1}ds,$$

where we have used the monotonicity of w in the last step. The proof is completed by noting that for $f, g \in H_w^0$ we have

$$\left|f(x)\int_0^x g(s)ds\right| \leq |f(x)|\|g\|\|I_x\| \to 0$$

as $x \to \infty$. □

Returning to our discussion of the long rate from Sect. 6.3.2, we note that for $f, g \in H_w^0$, the function F_{HJM} maps $\mathcal{L}_{\mathrm{HS}}(G, H_w^0)$ into the subspace H_w^0; i.e, for every $\sigma \in \mathcal{L}_{\mathrm{HS}}(G, H_w^0)$ we have $F_{\mathrm{HJM}}(\sigma)(\infty) = 0$. Hence, for HJM models on the space H_w, the long rate is constant.

6.4 Geometry of the Term Structure Dynamics

In this section we study some geometrical aspects to the evolution of the forward rate curve. As the properties studied in this chapter are almost sure properties of the models, we restrict ourselves to risk neutral probability structures. Consequently, the HJM models will be completely determined by the volatility σ since we can assume $\lambda = 0$. Moreover, for the sake of further convenience, we will drop the time dependence and work with a Markovian HJM model in a space F satisfying Assumption 6.1.

In this section, we study several problems relating the dynamics given by the HJM stochastic differential equations and their initial conditions.

- The first problem is of a stability nature. Given a set of initial forward curves satisfying a specific set of properties, we ask if the dynamics imposed by a volatility matrix σ are preserving these properties.
- The second problem is directly related to the factor models introduced in Sect. 2.1 of Chap. 2. Given a volatility structure σ, under which conditions will a k-factor representation in the sense of Chap. 2 exist?

6.4.1 The Consistency Problem

In this section we try to reconcile the point of view of Chap. 1 with the HJM framework presented in this chapter. Throughout this subsection, we shall also assume that the Wiener process is finite dimensional, and as usual, we denote by d its dimension. Under these conditions, the d-dimensional vector of volatilities is given by a deterministic function $f \hookrightarrow \sigma(f)$ of the forward rate curve f. Recall that the forward rate curve at a given time is not directly observable on the market. Rather, the forward rate curve is estimated from the market data, typically price quotes of coupon bonds and swap contracts. As we saw in Chap. 1, there are two main approaches: either the data are fit to a parametric family of curves such as the Nelson–Siegel family, or the forward rate curve is estimated by a nonparametric method, such as smoothing splines.

On the other hand, and independently of the choice of the initial forward curve, we model the dynamics of the forward rates by an HJM model for the purpose of pricing and hedging. In this section, we investigate whether these two modeling perspectives, the static parametric modeling of the initial forward rate curve and the choice of no arbitrage dynamics for its evolution, are compatible.

Let $f : \mathbb{R}_+ \times \mathbb{R}^k \to \mathbb{R}$ be a smooth function which we think of as a family of parameterized forward rate curves. That is, for a fixed $z \in \mathbb{R}^k$, we view $x \hookrightarrow f(x, z)$ as a forward rate curve which corresponds to the parameter z. The classical example of the Nelson–Siegel family was introduced in Chap. 1. For the sake of completeness we recall that this family is given by

$$f_{\mathrm{NS}}(x, z) = z_1 + (z_2 + z_3 x)e^{-x z_4}$$

with the four-dimensional parameter $z = (z_1, z_2, z_3, z_4)$.

Let us now fix $(x, t) \in \mathbb{R}_+ \times \mathbb{R}_+$ and let us assume that at each time $t \geq 0$, the forward rate $f_t(x)$ is given by $f_t(x) = f(x, Z_t)$ where Z_t is an $\mathcal{F}_t$-measurable random variable taking values in $\mathbb{R}^k$. In the spirit of Chap. 2, we consider the simplest model for the process $\{Z_t\}_{t\geq 0}$, and we assume that it is the solution to an SDE of the form

$$dZ_t = \mu^{(Z)}(Z_t)dt + \sigma^{(Z)}(Z_t)dW_t$$

under a martingale measure $\mathbb{Q}$. We may ask, then, for what functions $\mu^{(Z)}$ and $\sigma^{(Z)}$ are the dynamics of the forward rates $f_t(x)$ consistent with the HJM drift condition? The definite answer is given in the following theorem of Filipović:

Theorem 6.3. *For the dynamics of the system of forward curves $f_t(x) = f(x, Z_t)$ to satisfy the HJM drift condition $\alpha = F_{\mathrm{HJM}}(\sigma)$ (or in other words (6.2) with $\lambda = 0$), the following must hold identically in x and z:*

$$\frac{\partial}{\partial x}f(x,z) + \sum_{i,j=1}^{k} \sigma_i^{(Z)}(z)\sigma_j^{(Z)}(z)^* \frac{\partial}{\partial z_i}f(x,z)\int_0^x \frac{\partial}{\partial z_j}f(s,z)ds$$
$$= \sum_{i=1}^{k} \frac{\partial}{\partial z_i}f(x,z)\mu_i^{(Z)}(z) + \frac{1}{2}\sum_{i,j=1}^{k} \frac{\partial^2}{\partial z_i \partial z_j}f(x,z)\sigma_i^{(Z)}(z)\sigma_j^{(Z)}(z)^*.$$

An interesting consequence is that the Nelson–Siegel family is, in fact, inconsistent with every nontrivial HJM model. That is, if we knew that the forward rate at every time $t \geq 0$ was given by

$$f_t(x) = Z_t^{(1)} + (Z_t^{(2)} + Z_t^{(3)}x)\exp(-xZ_t^{(4)})$$

for some diffusion $\{Z_t^{(1)}, \ldots, Z_t^{(4)}\}_{t\geq 0}$, then the forward rates would necessarily be a deterministic function of the time 0 value of the process. To be specific we would have:

$$f_t(x) = Z_0^{(1)} + (Z_0^{(2)} + Z_0^{(3)}(x+t))\exp(-(x+t)Z_0^{(4)}),$$

in other words, the dynamics of the forward rates are given by the left-shift semigroup $f_t = S_t f_0$.

To be fair, central bankers do not believe that for every moment of time the forward curve is given by an element of the Nelson–Siegel family. Recall from the discussion of Chap. 1 that they solve an optimization problem to find the member of the family that best fits observed prices. Therefore, Nelson–Siegel and Svensson curves are regarded as approximations that capture some economically meaningful features of the observed forward rates, such as the location of dips and humps. However, it is important to realize that they fail to be consistent with the no-arbitrage principle.

6.4.2 Finite Dimensional Realizations

We now study the question of existence of finite-dimensional realizations of an infinite-dimensional HJM model. Using the terminology introduced in Chap. 2, this problem can be viewed as the search for a factor model giving the HJM model we start from. In other words, given a risk neutral HJM model determined by a volatility structure σ on F, we seek conditions under which there exists a smooth function $f^{(Z)} : \mathbb{R}_+ \times \mathbb{R}^k \to \mathbb{R}$ and a k-dimensional diffusion $\{Z_t\}_t$ given by an SDE of the form

$$dZ_t = \mu^{(Z)}(Z_t)dt + \sigma^{(Z)}(Z_t)dW_t$$

such that $f_t(x) = f^{(Z)}(x, Z_t)$ for all t almost surely. As explained earlier, the random variables $Z_t^{(1)}, \ldots, Z_t^{(k)}$ can be thought of as economic factors, but they can also include benchmark forward rates. As we can see, the finite dimensional realization problem asks whether a given HJM model can be realized from a finite factor model, such as those introduced in Chap. 2.

Before tackling the problem in its full generality, we first consider the particular case of a Gauss-Markov HJM model for which we can give an explicit solution. So we fix a deterministic volatility structure $\sigma \in \mathcal{L}_{\mathrm{HS}}(G, F)$ and we assume that

$$f_t = S_t f_0 + \int_0^t S_{t-s} F_{\mathrm{HJM}}(\sigma) ds + \int_0^t S_{t-s} \sigma dW_s.$$

The model admits a finite-dimensional realization if and only if there exist a smooth F-valued function f such that

$$f^{(Z)}(Z_t) = f_t$$

for some k-dimensional diffusion $\{Z_t\}_{t \geq 0}$. If we let the first factor $Z_t^{(1)} = t$ be the running time, we can rewrite the above equation as

$$\bar{f}^{(Z)}(Z_t) = \int_0^t S_{t-s} \sigma dW_s \tag{6.8}$$

where

$$\bar{f}^{(Z)}(z) = f^{(Z)}(z) - S_{z^{(1)}} f_0 + \int_0^{z^{(1)}} S_{z^{(1)}-s} \alpha ds.$$

Since the right-hand side of Eq. (6.8) is mean zero and Gaussian, we study the case where $\bar{f}^{(Z)}$ is linear in $z^{(2)}, \dots, z^{(k)}$ and the diffusion $\{(Z_t^{(2)}, \dots, Z_t^{(k)})\}_{t \geq 0}$ is mean zero and Gaussian.

A direct calculation reveals that Eq. (6.8) implies that the range of the operator σ must be contained in a finite dimensional invariant subspace for the semigroup $\{S_t\}_{t \geq 0}$. Since elements of F are continuous by assumption, the only such subspaces are those spanned by quasi-exponential functions, where an element $g \in F$ is called *quasi-exponential* if g can be written as a finite linear combination of functions of the form

$$x \hookrightarrow x^n e^{-\beta x} \cos(\omega x) \quad \text{and} \quad x \hookrightarrow x^n e^{-\beta x} \sin(\omega x)$$

for a natural number n and real numbers β and ω. We record this result as a proposition:

Proposition 6.5. *A Gauss-Markov HJM model admits a realization as an affine function of a finite-dimensional Gaussian diffusion if and only if there is a finite number of quasi-exponential functions $g_1, \dots, g_d \in F$ such that*

$$\mathrm{range}(\sigma) \subset \mathrm{span}\{g_i;\ i = 1, \dots, d\}.$$

As an example, consider a scalar volatility $\sigma(x) = \sigma_0 e^{-\beta x}$. Then the forward rates are given by

$$f_t(x) = f_0(t+x) + \frac{\sigma_0^2}{\beta^2} e^{-\beta x}(1-e^{-\beta t})(1-e^{-\beta x}(1+e^{\beta t})/2) + \int_0^t \sigma_0 e^{-\beta(t-s+x)} dw_s.$$

Letting $x = 0$ in the above equation, we see that the short rate $r_t = f_t(0)$ is given by

$$r_t = f_0(t) + \frac{\sigma_0^2}{2\beta^2}(1 - e^{-\beta t})^2 + \int_0^t \sigma_0 e^{-\beta(t-s)} dw_s.$$

which can be expressed in differential form:

$$dr_t = (\alpha_t - \beta r_t)dt + \sigma_0 dw_t$$

where

$$\alpha_t = f_0'(t) + \beta f_0(t) + \frac{\sigma_0^2}{\beta}(1 - e^{-\beta t}).$$

For this example, the a priori infinite dimensional HJM model can be realized as a two-factor homogeneous Markov model with $Z_t^{(1)} = t$ and $Z_t^{(2)} = r_t$. In fact, this is just the Vasicek–Hull–White short rate model.

We should note in the Gauss-Markov case that the existence of a finite dimensional realization imposes a strong regularity condition on the possible forward rate curves the model can produce. Indeed, the function $x \hookrightarrow f_t(x)$ for $t \geq 0$ is as smooth as the initial forward curve $x \hookrightarrow f_0(x)$. This is because the quasi-exponential functions are real analytic. In fact, in the general Markovian case, it turns out that finite dimensional realizations are smooth.

Indeed, the existence of a finite dimensional realization of an HJM model at point $f_0 \in F$ is essentially equivalent to the existence of a finite dimensional manifold $M \subset F$ containing f_0 which is invariant for the HJM evolution. It is proven by Filipović and Teichmann in [64] that such invariant manifolds are contained in the set $\mathcal{D}(A^\infty) = \cap_{i=1}^\infty \mathcal{D}(A^i)$. So finite realizations are given by C^∞ vectors of the operator A in F. But since A is a differential operator, these C^∞ vectors are necessarily infinitely differentiable functions.

We now consider the finite dimensional realization problem for more general Markovian (time-homogeneous) HJM models. We fix a deterministic function $f \hookrightarrow \sigma(f)$ which we assume to be infinitely Fréchet differentiable. Again, we define $f \hookrightarrow \alpha(f)$ by

$$\alpha(f) = F_{\text{HJM}} \circ \sigma(f).$$

We will further assume that the ranges of σ and α are contained in the set $\mathcal{D}(A^\infty)$. As indicated by the title of this section, both the results and the tools used to prove them are of a geometric nature, and we need to introduce concepts and definitions which were not needed so far. The first of them is the concept of Lie bracket of vector fields. For the purpose of this book, a C^n-vector field on an open subset of F, say U, is an n-times Fréchet differentiable function from U into F. Most of the vector fields used in this book are C^∞, and the reader should assume a C^∞ vector field when no mention is made of the order of differentiability. Moreover, it will be intended that the vector field is defined on the full space whenever the open set U is not mentioned explicitly.

Definition 6.3. *The Lie bracket of two smooth vector fields G and F on U is defined as the function $[G, H] : U \to F$ given by*

$$[G, H](f) = G'(f)H(f) - H'(f)G(f).$$

The Lie algebra generated by a family of vector fields $G_1, \ldots, G_n$ on U is the smallest linear space of vector fields on U containing $G_1, \ldots, G_n$ which is closed under the Lie bracket. We denote this Lie algebra by

$$\{G_1, \ldots, G_n\}_{LA}.$$

Recall that we use the notation $'$ to denote the Fréchet derivative of a function. A finite dimensional realization depends on the volatility $\sigma : F \to \mathcal{L}_{\mathrm{HS}}(\mathbb{R}^d, F)$ and the initial forward curve $f_0 \in F$. We are interested here in those realizations which are stable under perturbation of the initial curve in a sense to be made precise later. We will say that a volatility σ has a *generic* finite dimensional realization at f_0 if there exists a finite dimensional realization at each point of a neighborhood of f_0.

Notice that in general A is not a smooth vector field on F, so that $\mathcal{D}(A^\infty)$ is usually a small subset of F. Therefore, for convenience in stating the following Frobenius type theorem, we impose one additional, and quite strong, assumption: The derivative operator $A = d/dx$ is bounded on F. We comment on the meaning of this restrictive assumption later in this subsection.

Theorem 6.4. *The HJM equation generically admits a finite dimensional realization if and only if the Lie algebra*

$$\{\mu, \sigma_1, \ldots, \sigma_d\}_{LA}$$

is finite dimensional in a neighborhood of $f_0 \in F$, where

$$\mu(f) = \frac{\partial}{\partial x} f + \alpha(f) - \frac{1}{2} \sum_{i=1}^{d} \sigma_i'(f)\sigma_i(f)$$

is the drift term of the HJM equation written in Stratonvich form and $\sigma_i(f) = \sigma(f)g_i$ for some CONS $\{g_i\}_i$ of G.

Recall that the Stratonvich integral $\int_0^t X_s \circ dY_s$ is defined for semimartingales X and Y by the formula

$$\int_0^t X_s \circ dY_s = \int_0^t X_s dY_s + \frac{1}{2} \ll X, Y \gg_t$$

where the first integral on the right is an Itô integral and where $\ll X, Y \gg_t$ denotes the quadratic co-variation of the processes X and Y. The reason for the sudden appearance of the Stratonvich integral is that Itô's formula can be rewritten in the form

$$dg(X_t) = g'(X_t) \circ dX_t$$

for smooth g. In particular, the chain rule for Stratonvich calculus is exactly the same as the ordinary chain rule.

The assumption that A is bounded is too strong. Note that since

$$|f^{(n)}(0)| = |\langle \delta_0, A^n f\rangle| \leq \|\delta_0\| \|f\| \|A\|^n$$

every function $f \in F$ has a Taylor series

$$f(x) = \sum_{n=0}^{\infty} \frac{f^{(n)}(0)}{n!} x^n$$

which converges for all $x \geq 0$. That is to say, for A to be a bounded operator on F, the space F has to consist of real analytic functions. However, the forward curves produced by the CIR model cannot be extended to entire functions due to the presence of complex poles.

Fortunately, the above theorem has been extended by Filipović and Teichmann to the more general case of unbounded operators A on F. The key idea is to restrict attention to the Fréchet space $\mathcal{D}(A^\infty)$ of C^∞ vectors of A. Because of the lack of a norm topology, the differential calculus is carried out in the framework of so-called convenient analysis. See the Notes & Complements for references.

The surprising consequence of this theory is that all generic finite dimensional realizations come from affine factor models:

Theorem 6.5. *Suppose that there are p linearly independent functionals $\ell_1, \dots, \ell_p \in F^*$ such that $\sigma(f) = \sigma(\ell_1(f), \dots, \ell_p(f))$. Let U be an open subset of the Fréchet space $\mathcal{D}(A^\infty)$. If there exists a generic finite dimensional realization at $f_0 \in U$ then the forward curve is of the affine form:*

$$f_t = A + \sum_{i=1}^{k} Z_t^{(i)} B_i$$

for a k-dimensional diffusion $\{Z_t\}_{t \geq 0}$.

We now see why the Hull–White extensions of the Vasicek and Cox–Ingersoll–Ross models behave so nicely: They are essentially the only generic short rate models! All other short rate models are "accidental" in the following sense: Suppose that we start with a short rate model that is not affine. By solving the appropriate PDE, we find the function $f^{(r)}$ which determines the forward rates given the short rate. By applying Itô's formula to $f^{(r)}(x, r_t)$ we obtain an HJM equation for the forward rate evolution. By construction, this HJM model admits a finite dimensional realization at $f_0 = f^{(r)}(r_0)$. However, the HJM model does *not* admit a finite dimensional realization at some initial point f in every neighborhood of f_0!

6.5 Generalized Bond Portfolios

In this section we consider a generalized notion of bond portfolios. In particular, we allow our investor to own bonds of infinite number of maturities.

Let us recall one of the original motivations for introducing HJM models driven by an infinite dimensional Wiener process: In a finite rank HJM model driven by a d-dimensional Wiener process, every square-integrable claim can be replicated by a strategy of holding bonds maturing at the d dates $T_1, \ldots, T_d$ fixed a-priori and independently of the claim. For instance, with a three-factor HJM model, it is possible to perfectly hedge a call option on a bond of maturity five years with a portfolio of bonds of maturity twenty, twenty-five, and thirty years. This result is counter-intuitive and contrary to market practice. Indeed, there seems to be a notion of "maturity specific risk" not captured by finite rank HJM models.

There is a tremendous amount of redundancy in a finite factor model if the investor is allowed to trade in bonds of any maturity. In particular, there is no uniqueness of hedging strategies. In Sect. 6.5.3 below we show that if the dynamics of the bond prices are driven by an infinite dimensional Wiener process we can find conditions on the model ensuring that a given hedging strategy is unique. Unfortunately, the usual notions of hedging become more complicated in infinite dimensions.

As usual, let $P_t(x)$ denote the price at time $t \geq 0$ of a zero-coupon bond with time to maturity $x \geq 0$. If at time $t \geq 0$, the investor owns $c_1, \ldots, c_N$ units of the bonds with time to maturities $x_1, \ldots x_N$ respectively, then his wealth is simply given by

$$X_t = \sum_{i=1}^{N} c_i P_t(x_i).$$

We can rewrite the wealth in the more suggestive form

$$X_t = \left(\sum_{i=1}^{N} c_i \delta_{x_i} \right) (P_t)$$

where as before δ_x denotes the evaluation functional $\delta_x(f) = f(x)$. We may say a portfolio is simple if it can be realized as a linear combination of a finite number of evaluation functionals. Alternatively, since the evaluation functional δ_x can be identified with the point mass $\delta_x(A) = \mathbf{1}_A(x)$, a simple portfolio is just an atomic measure on $\mathbb{R}_+$ with a finite number of atoms.

The generalized portfolios considered in this section, then, are limits of simple portfolios. We present a framework in which an investor may choose a portfolio, for instance, which may be realized as continuous finite measures on $\mathbb{R}_+$.

To facilitate the exposition we break from the HJM tradition and choose for the state variable the discounted bond price curve $\tilde{P}_t(\cdot)$ instead of the forward rate curve $f_t(\cdot)$.

This change of variables eases the analysis, although it is quite superficial in the sense that there is a one-to-one correspondence between discounted bond prices and instantaneous forward rates. Indeed, we can use the formulas

$$\tilde{P}_t(x) = \exp\left(-\int_0^t f_s(0)ds - \int_0^x f_t(s)ds\right)$$

and

$$f_t(x) = -\frac{\partial}{\partial x}\log(\tilde{P}_t(x)).$$

to go from discounted bond prices to forward rates and back.

6.5.1 Models of the Discounted Bond Price Curve

We now formulate a model of the discounted price dynamics. Paralleling the treatment of the abstract HJM forward rate models, we first choose our state space F of bond price curves.

Assumption 6.4. 1. The space F is a separable Hilbert space and the elements of F are continuous, real-valued functions on $\mathbb{R}_+$. We also assume that, for every $x \in \mathbb{R}_+$, the evaluation functional $\delta_x(f) = f(x)$ is a continuous linear function on F.
2. The semigroup $\{S_t\}_{t\geq 0}$ of left shift operators

$$(S_t f)(x) = f(t+x)$$

is strongly continuous.

As before, the infinitesimal generator of $\{S_t\}_{t\geq 0}$ will be denoted by A. We do not assume that A is bounded. Notice also that in order to make the typical element of F resemble a real bond price curve, we also can assume that

$$\lim_{x\to\infty} f(x) = 0$$

for all $f \in F$. However, this assumption is never needed in what follows.

Although F is a Hilbert space, and we can identify the elements of F and the dual space F^*, it is more natural to think of F as a space of functions and F^* a space of functionals. For this reason we insist in this section more than usual on the distinction between a space and its dual, and in particular the bracket $\langle\cdot,\cdot\rangle_F$ is reserved for the duality form $F^* \times F \to \mathbb{R}$.

Here and throughout this section, the stochastic processes are assumed to be defined on a complete probability space $(\Omega, \mathcal{F}, \mathbb{Q})$ supporting a Wiener process $\{W_t\}_{t\geq 0}$. Since the state space of the Wiener process is irrelevant, we can assume that $\{W_t\}_t$ is cylindrically defined on a separable Hilbert space G. Let $\{\mathcal{F}_t\}_{t\geq 0}$ be the augmentation of the filtration generated by the Wiener process, and we assume that sigma-field $\mathcal{F}$ is also generated by the Wiener process. Let $\mathcal{P}$ be the predictable sigma-field on $\mathbb{R}_+ \times \Omega$.

We now state the definition of what we mean by abstract risk-neutral discounted bond price model.

Definition 6.4. *A risk-neutral discounted bond price model on F is a measurable function σ from $(\mathbb{R}_+ \times \Omega \times F, \mathcal{P} \otimes \mathcal{B}_F)$ into $(\mathcal{L}_{\text{HS}}(G,F), \mathcal{B}_{\mathcal{L}_{\text{HS}}(G,F)})$, such that for all $(t,\omega,f) \in \mathbb{R}_+ \times \Omega \times F$ we have*

$$\sigma(t,\omega,f)^*\delta_0 = 0 \tag{6.9}$$

and such that there exists a non-empty set of initial conditions $\tilde{P}_0 \in F$ for which there exists a unique, continuous mild F-valued solution $\{\tilde{P}_t\}_{t\geq 0}$ of the evolution equation:

$$d\tilde{P}_t = A\tilde{P}_t dt + \sigma(t,\omega,\tilde{P}_t)dW_t \tag{6.10}$$

with the property that

$$\tilde{P}_t(x) > 0 \tag{6.11}$$

almost surely for all (t,x). Once the initial conditions $\tilde{P}_0 \in F$ is fixed, we use the abbreviation $\sigma_t = \sigma(t,\omega,\tilde{P}_t)$.

Fixing an initial condition $\tilde{P}_0 \in F$, the discounted bond prices satisfy the integral equation

$$\tilde{P}_t = S_t\tilde{P}_0 + \int_0^t S_{t-s}\sigma(s,\omega,\tilde{P}_s)dW_s$$

Given a model for the discounted bond prices, we have for every fixed $T > 0$ that the process $\{\tilde{P}_t(T-t)\}_{t\in[0,T]}$ is a continuous local martingale. Indeed, we have the calculation:

$$\begin{aligned}\tilde{P}_t(T-t) &= \langle S^*_{T-t}\delta_0, \tilde{P}_0\rangle_F \\ &= \tilde{P}_0(T) + \int_0^t \sigma^*_s S^*_{T-s}\delta_0 dW_s.\end{aligned}$$

The condition $\sigma^*(t,\omega,f)\delta_0 = 0$ stated as (6.9) implies that the volatility of the discounted bond vanishes at maturity. We have started with the discounted bond prices as our modeling primitive. Assuming that the discounted bond prices are positive, we can recover the bank account process by the formula

$$B_t = \tilde{P}_t(0)^{-1}$$

and the actual bond prices by:

$$P_t(x) = \tilde{P}_t(0)^{-1}\tilde{P}_t(x).$$

It is easy to see that a sufficient condition on the function σ which guarantees that the discounted bond prices are positive is

$$\|\sigma(t,\omega,f)^*\delta_x\|_G \leq C|f(x)|$$

for some constant $C > 0$.

6.5.2 Trading Strategies

In this section we consider the notion of trading strategy in the bond market model introduced in Definition 6.4. Let ϕ_t be an investor's vector of portfolio weights at time $t \geq 0$. Recall that for each $(t, \omega) \in \mathbb{R}_+ \times \Omega$, the discounted bond price vector $\tilde{P}_t$ is an element of the Hilbert space F; therefore, the undiscounted bond price vector $P_t = \tilde{P}_t(0)^{-1}\tilde{P}_t$ is also in F. Since we would like to compute the investor's wealth by the formula

$$X_t = \langle \phi_t, P_t \rangle_F,$$

we assume that the vector ϕ_t is valued in the dual space F^*.

By assumption, the dual space F^* is a space of distributions which contains the finite measures on $\mathbb{R}_+$. In particular, we can approximate ϕ_t in F^* by a linear combination

$$\phi_t \approx \sum_{i=1}^{N} c_i \delta_{x_i}$$

corresponding to the simple portfolio of holding c_i units of the bond with time to maturity x_i for $i = 1, \ldots, N$.

If an investor is holding the portfolio ϕ_t at time $t \geq 0$, then the discounted wealth is given by

$$\tilde{X}_t = \langle \phi_t, \tilde{P}_t \rangle_F.$$

Since the process $\{\tilde{P}_t\}_{t \geq 0}$ is generally not a semi-martingale, a modicum of care is needed in formulating the self-financing condition.

In order to keep track of notation, and in particular to pass back from Musiela's relative maturity notation to the absolute maturity notation, we temporarily introduce the semigroup $\{S_{-t}\}_{t \geq 0}$ of right shifts. The crucial property we need is that for every $t \geq 0$ the operator S_{-t} is the right inverse of S_t; that is, we have $S_t \circ S_{-t} = I$ where I is the identity on F. This is accomplished by letting $(S_{-t}f)(x) = f(x - t)$ for all $x \geq t$. As we will see below, we will not have to evaluate the expression $(S_{-t}f)(x)$ for $t > x$. If for instance the elements of F are real analytic functions, then each $f \in F$ can be extended in a unique way to a function on all of $\mathbb{R}$, in which case the expression $(S_{-t}f)(x) = f(x - t)$ is unambiguous. However, it is unnecessary to assume so much smoothness. In order to be specific, we might as well let $(S_{-t}f)(x) = f(0)$ for $t > x$. However, we should keep in mind the fact that the collection $\{S_t\}_{t \in \mathbb{R}}$ of operators defined in this way is generally not a group since $S_{-t} \circ S_t \neq I$.

Suppose for the moment that S_{-t} is bounded on F for $t \geq 0$. Then the investor wealth is given by

$$\tilde{X}_t = \langle S_t^* \phi_t, S_{-t}\tilde{P}_t \rangle_F.$$

To make sense of the above expression, suppose $\{S_{-t}\}_{t\geq 0}$ is strongly continuous on F, and note that the process $\{S_{-t}\tilde{P}_t\}_{t\geq 0}$ satisfies the equation

$$\begin{aligned} S_{-t}\tilde{P}_t &= S_{-t}\left(S_t\tilde{P}_0 + \int_0^t S_{t-s}\sigma_s dW_s\right) \\ &= \tilde{P}_0 + \int_0^t S_{-s}\sigma_s dW_s. \end{aligned}$$

Formally, the self-financing condition should read

$$d\tilde{X}_t = \langle S_t^*\phi_t, dS_{-t}\tilde{P}_t\rangle_F.$$

and hence, the discounted wealth process should have self-financing dynamics given by

$$d\tilde{X}_t = \sigma_t^*\phi_t dW_t.$$

Now that we have our desired formula, we can dispense with any assumptions about the semigroup $\{S_{-t}\}_{t\geq 0}$. To formalize this discussion, we give the following definition:

Definition 6.5. *A strategy is an adapted F^*-valued process $\{\phi_t\}_{t\geq 0}$ such that*

$$\mathbb{E}\left\{\int_0^t \|\sigma(s,\omega,\tilde{P}_s)^*\phi_s\|_G^2 ds\right\} < +\infty$$

almost surely for all $t \geq 0$. A strategy $\{\phi_t\}_{t\geq 0}$ is self-financing if there exists a constant $X_0 \in \mathbb{R}$ such that

$$\langle \phi_t, \tilde{P}_t\rangle_F - \int_0^t \sigma_s^*\phi_s dW_t = X_0$$

for all $t \geq 0$ almost surely. If $\{\phi_t\}_{t\geq 0}$ is self-financing we let

$$\tilde{X}_t^\phi = \langle \phi_t, \tilde{P}_t\rangle_F = X_0 + \int_0^t \sigma_s^*\phi_s dW_t$$

denote the corresponding discounted wealth at time $t \geq 0$.

Here are a few examples of self-financing strategies.

- We have already encountered one example of a self-financing strategy given by $\phi_t = \delta_{T-t}$ for some fixed $T > 0$. This strategy corresponds to buying and holding the bond with maturity T. In this case, the constant X_0 is given by the initial bond price $\tilde{P}_0(T) = P_0(T)$.
- Another example of a self-financing strategy is given by $\phi_t = B_t\delta_0 = \tilde{P}_t(0)^{-1}\delta_0$. This strategy corresponds to an investor always holding the just-maturing bond and reinvesting the instantaneous return. The discounted wealth is constant $\tilde{X}_t = 1$ for all $t \geq 0$, and hence the undiscounted wealth is given by the bank account $X_t = B_t$.

- We can use the last example of the bank account to construct from any strategy $\{\phi_t\}_{t\geq 0}$ a self-financing strategy $\{\varphi_t\}_{t\geq 0}$ by fixing an initial wealth X_0 and letting

$$\varphi_t = \phi_t + \psi_t B_t \delta_0$$

where the scalar random variable

$$\psi_t = X_0 + \int_0^t \sigma_s^* \phi_s dW_s - \langle \phi_t, \tilde{P}_t \rangle_F$$

corresponds to the portion of the investor's wealth held in the bank account.

6.5.3 Uniqueness of Hedging Strategies

In this section, we reap the benefits of working with an infinite dimensional Wiener process $\{W_t\}_{t\geq 0}$. Recall from Chap. 2 that if an HJM model is of finite rank d then all contingent claims can be hedged with the bank account and a strategy consisting of bonds of d arbitrarily chosen maturities. We claimed that passing to infinite ranked models may introduce some notion of maturity specific risk. We now quantify this assertion by finding a condition such that hedging strategies are unique.

Note that by Definition 6.4 of the discounted bond price model we necessarily have for all $(t, \omega, f) \in \mathbb{R}_+ \times \Omega \times F$ that

$$\text{range}(\sigma(t, \omega, f)) \subset \{g \in F;\ g(0) = 0\}$$

or equivalently

$$\ker(\sigma(t, \omega, f)^*) \supset \text{span}\{\delta_0\}.$$

If we insist that these inclusions be equalities, we have the following uniqueness result:

Proposition 6.6. *Suppose that for all $(t, \omega, f) \in \mathbb{R}_+ \times \Omega \times F)$ we have:*

$$\ker(\sigma(t, \omega, f)^*) = \text{span}\{\delta_0\}. \tag{6.12}$$

If there is a deterministic time $T \geq 0$ such that the self-financing strategies $\{\phi_t^1\}_{t\in[0,T]}$ and $\{\phi_t^2\}_{t\in[0,T]}$ are such that the discounted wealths coincide in the end, i.e. satisfy $\tilde{X}_T^{\phi^1} = \tilde{X}_T^{\phi^2}$, then $\phi_t^1 = \phi_t^2$ for almost all $(t, \omega) \in [0, T] \times \Omega$.

Proof. Let $\psi_t = \phi_t^1 - \phi_t^2$ define the self-financing strategy such that $\tilde{X}_T^{\psi} = 0$. We have $\mathbb{E}\{\tilde{X}_T^{\psi}\} = X_0^{\psi} = 0$ and $\int_0^T \sigma_t \psi_t dW_t = 0$ almost surely and hence

$$\mathbb{E}\left\{\int_0^T \|\sigma_t^* \psi_t\|_G^2 dt\right\} = 0.$$

We conclude that for almost all (t, ω) the functional ψ_t is a scalar multiple of δ_0. Since $\{\psi_t\}_{t \in [0,T]}$ is self-financing we have $\langle \psi_t, \tilde{P}_t \rangle_F = 0$ for almost all (t, ω). Letting $\psi_t = c_t \delta_0$ for a scalar random variable c_t, we must have $c_t \tilde{P}_t(0) = 0$. But by assumption $\tilde{P}_t(0) > 0$ so we conclude that $c_t = 0$ and the result is proven. □

Only with infinite dimensional G can we hope to satisfy the conditions of the above proposition.

6.5.4 Approximate Completeness of the Bond Market

In this section, we explore an annoying complication of HJM models driven by infinite-dimensional Wiener processes. Unlike the finite dimensional case, the bond market model is incomplete – even if there exists a unique martingale measure! This might come as a surprise since the Black–Scholes–Merton model taught that a model is complete if, roughly speaking, there exists as many traded assets as independent Wiener processes. We therefore might expect that the bond market model is complete since we have allowed distribution-valued trading portfolios which might consist of a continuum of bonds.

However, there are square-integrable contingent claims which cannot be replicated by F^*-valued trading strategies. Parroting the calculation from Chap. 2, let $\tilde{\xi}$ be a square discounted integrable claim and suppose that we could find a self-financing strategy $\{\phi_t\}_{t \geq 0}$ such that:

$$\begin{aligned} \tilde{\xi} &= \tilde{X}_T^{\phi} \\ &= \tilde{X}_0^{\phi} + \int_0^T \sigma_s^* \phi_s dW_s. \end{aligned}$$

But recall that the martingale representation theorem states that there exists a predictable G-valued process $\{\alpha_t\}_{t \in [0,T]}$ such that $\int_0^T \|\alpha_s\|_G^2 ds < +\infty$ and

$$\tilde{\xi} = \mathbb{E}\{\tilde{\xi}\} + \int_0^T \alpha_s dW_s. \tag{6.13}$$

Note that we identify G and its dual G^* even though we insisted that we should not identify the state space F and its dual F^*. Thus, in order to calculate a hedging portfolio at time $t \in [0, T]$ we need only compute $\phi_t = \sigma_t^{*-1} \alpha_t$. But by assumption the operator σ_t is Hilbert–Schmidt almost surely. Since G is infinite dimensional, the inverse σ_t^{*-1} is unbounded, and at this level of generality there is no guarantee that α_t is in its domain for any $t \in [0, T]$. Thus restricting the portfolio to be in the space F^* for all $t \in [0, T]$ is insufficient to replicate every square-integrable contingent claim.

Despite the above bad news, if for all (t, ω, f) the range of the operator $\sigma(t, \omega, f)^*$ is dense in G, or equivalently the kernel of the operator $\sigma(t, \omega, f)$

is $\{0\}$, then the market is *approximately complete*. That is, for every reasonable claim ξ and $\epsilon > 0$, there exists a strategy $\{\phi_t^{(\epsilon)}\}_{t\in[0,T]}$ which approximately hedges ξ in the mean square sense:

$$\mathbb{E}\left\{\left(\mathbb{E}\{\tilde{\xi}\} + \int_0^T \sigma_t^* \phi_t^{(\epsilon)} dW_t - \tilde{\xi}\right)^2\right\} < \epsilon.$$

6.5.5 Hedging Strategies for Lipschitz Claims

In this section we find explicit hedging strategies for an important class of contingent claims, and we characterize their properties. We are interested in European bond options: options which mature at fixed future date $T > 0$, and are written on bonds with fixed maturities $T < T_1 < \cdots < T_n$. The payout of such an option is given by

$$\begin{aligned}\xi &= g(P(T,T_1), \ldots, P(T,T_n)) \\ &= g(P_T(T_1 - T), \ldots, P_T(T_n - T))\end{aligned}$$

at maturity, for some function $g : \mathbb{R}^n \to \mathbb{R}$.

We have in mind the example of a call option with expiration T and strike K, on a bond with maturity $T_1 > T$. The payout function in this case is $g(x) = (x - K)^+$.

Consider the discounted contingent claim $\tilde{\xi} = B_T^{-1}\xi$. It can be written in the form

$$\begin{aligned}\tilde{\xi} &= \tilde{P}_T(0) g\left(\frac{\tilde{P}_T(T_1 - T)}{\tilde{P}_T(0)}, \ldots, \frac{\tilde{P}_T(T_n - T)}{\tilde{P}_T(0)}\right) \\ &= \tilde{g}(\tilde{P}_T)\end{aligned}$$

where the function $\tilde{g} : F \to \mathbb{R}$ is defined by

$$\tilde{g}(f) = f(0) g\left(\frac{f(T_1 - T)}{f(0)}, \ldots, \frac{f(T_n - T)}{f(0)}\right).$$

It is this function $\tilde{g}$ that we will call the discounted payout function.

For the remainder of this section we make the following standing assumption:

Assumption 6.5. The contingent claim is European with expiration T and discounted payout given by $\tilde{\xi} = \tilde{g}(\tilde{P}_T)$. The payout function $\tilde{g} : F \to \mathbb{R}$ satisfies the Lipschitz bound

$$|\tilde{g}(f_1) - \tilde{g}(f_2)| \leq C\|f_1 - f_2\|_F \tag{6.14}$$

for all $f_1, f_2 \in F$ and some constant $C > 0$.

We note that the Lipschitz assumption is reasonable. For example, the call option on a bond with maturity $T_1 > T$ has a discounted payout of the form

$$\begin{aligned}\tilde{g}(f) &= f(0)\left(\frac{f(T_1 - T)}{f(0)} - K\right)^+ \\ &= (f(T_1 - T) - f(0))^+\end{aligned}$$

for $f(0) > 0$, which is Lipschitz since the evaluation functionals δ_0 and $\delta_{T_1 - T}$ are bounded.

For the remainder of this section we work in a Markovian setting. The dynamics of the discounted bond prices will be given by an abstract discounted bond price model which satisfies

$$\sigma(t, \omega, f) = \bar{\sigma}(t, f)$$

for some function $\bar{\sigma} : \mathbb{R}_+ \times F \to \mathcal{L}_{\mathrm{HS}}(G, F)$. To lighten the notation, we drop the bar and let $\sigma = \bar{\sigma}$. We list here the relevant assumptions on $\sigma(\cdot, \cdot)$.

Assumption 6.6. Let $\sigma(\cdot, \cdot) : \mathbb{R}_+ \times F \to \mathcal{L}_{\mathrm{HS}}(H, F)$ be such that $\sigma(\cdot, f)$ is continuous for all $f \in F$, and satisfies the Lipschitz bound

$$\|\sigma(t, f_1) - \sigma(t, f_2)\|_{\mathcal{L}_{\mathrm{HS}}(G,F)} \leq C\|f_1 - f_2\|_F \tag{6.15}$$

for all $t \geq 0$, $f_1, f_2 \in F$ and some $C > 0$.

In Chap. 5 we proved that mild solutions to Markovian evolution equations with Lipschitz coefficients are Malliavin differentiable. In particular, we have $\tilde{P}_T \in \mathbb{H}(F)$. Furthermore, there is a strong random operator valued process $\{Y_{t,T}\}_{t \in [0,T]}$ such that the Malliavin derivative of the discounted bond price can be written in the form

$$D_t \tilde{P}_T = Y_{t,T} \sigma_t.$$

We assumed that $\tilde{g}$ is Lipschitz, so we have the chain rule

$$D_t \tilde{g}(\tilde{P}_T) = \nabla \tilde{g}(\tilde{P}_T) D_t \tilde{P}_T.$$

We can identify our candidate strategy from the Clark–Ocone formula:

$$\begin{aligned}\tilde{g}(\tilde{P}_T) &= \mathbb{E}\{\tilde{g}(\tilde{P}_T)\} + \int_0^T \mathbb{E}\{D_t \tilde{g}(\tilde{P}_T) | \mathcal{F}_t\} dW_t \\ &= \mathbb{E}\{\tilde{g}(\tilde{P}_T)\} + \int_0^T \sigma_t^* \mathbb{E}\{Y_{t,T}^* \nabla \tilde{g}(\tilde{P}_T) | \mathcal{F}_t\} dW_t.\end{aligned}$$

This preamble leads to the following desirable result.

Proposition 6.7. *For each $t \in [0,T]$, let $\varphi_t = \phi_t + B_t\psi_t\delta_0$ where*

$$\phi_t = \mathbb{E}\{Y_{t,T}^* \nabla\tilde{g}(\tilde{P}_T)|\mathcal{F}_t\} \tag{6.16}$$

and

$$\psi_t = \mathbb{E}\{\tilde{g}(\tilde{P}_T)|\mathcal{F}_t\} - \langle \phi_t, \tilde{P}_t\rangle_F$$

and where $\{Y_{s,t}\}_{0\le s\le t}$ is a strong operator (from F into F) solution to the equation

$$dY_{s,t} = AY_{s,t}dt + \nabla\sigma_t Y_{s,t}\cdot dW_t$$

with $Y_{s,s} = I$ for all $s \ge 0$. Then the process $\{\varphi_t\}_{t\in[0,T]}$ is a self-financing strategy which replicates the claim ξ.

Note that the portfolio ϕ_t is defined by a conditional expectation of a vector valued random variable. This expectation must be interpreted in the weak sense, not in the Bochner sense.

Proof. By the formal calculation above, we only need to show that ϕ_t is valued in F^* for almost all $t \in [0,T]$. In fact, we have that ϕ_t is bounded in F^* uniformly in $t \in [0,T]$ almost surely. Indeed, we have

$$\begin{aligned}\|\phi_t\|_{F^*} &= \sup_{\|f\|_F\le 1} \mathbb{E}\{\langle \nabla\tilde{g}(\tilde{P}_T), Y_{t,T}f\rangle_F|\mathcal{F}_t\} \\ &\le \sup_{\|x\|_F\le 1} C\mathbb{E}\{\|Y_{t,T}f\|_F^2|\mathcal{F}_t\}^{1/2}\end{aligned}$$

by the Lipschitz bound (6.14) and the bound for the moments of mild solutions of SDEs derived in the previous chapter. □

Revisiting the motivating example of Chap. 2, for any contingent claim maturing at time T, we denote by $T_{\max} > T$ the longest maturity of the bonds underlying the claim. The following theorem shows that under the appropriate assumptions in the case of infinite factor HJM models, the bonds in the hedging strategy for this claim have maturities less than or equal to $T_{\max}$. This intuitively appealing result is inspired by the local HJM models of the type

$$df_t = (Af_t + F_{\mathrm{HJM}}\circ\tau(f_t))dt + \tau(f_t)dW_t$$

where for all $x \ge 0$ the volatility function is such that

$$\tau(f)^*\delta_x = \kappa(x, f(x))$$

for a deterministic function $\kappa : \mathbb{R}_+ \times \mathbb{R} \to G$. Translating the forward rate model into a discounted bond price model we have

$$d\tilde{P}_t = A\tilde{P}_t dt + \sigma(\tilde{P}_t)dW_t$$

where

$$\sigma(f)^*\delta_x = f(x)\int_0^x \kappa\left(s, -\frac{f'(s)}{f(s)}\right) ds.$$

Note that for these models, the volatility of the discounted bond price with time to maturity x is a function only of the discounted prices of bonds with times to maturity less than or equal to x.

Theorem 6.6. *Let us assume that:*

- *there exists a constant $x_{\max} \geq 0$ such that the discounted payout function $\tilde{g}(\cdot) : F \to \mathbb{R}$ satisfies*

$$\tilde{g}(f_1) = \tilde{g}(f_2) \tag{6.17}$$

whenever $f_1(x) = f_2(x)$ for all $x \in [0, x_{\max}]$,

- *the volatility function $\sigma(\cdot,\cdot) : \mathbb{R}_+ \times F \to \mathcal{L}_{\mathrm{HS}}(G, F)$ is such that for all $t \geq 0$ and $x \geq 0$,*

$$\sigma(t, f_1)^*\delta_x = \sigma(t, f_2)^*\delta_x \tag{6.18}$$

whenever $f_1(y) = f_2(y)$ for all $y \in [0, x]$,

- *and for all $f \in F$ and $t \geq 0$ we have*

$$\ker(\sigma(t, f)^*) = \mathrm{span}\{\delta_0\}.$$

Then, the unique strategy $\{\varphi_t\}_{t\in[0,T]}$ given by Eq. (6.16) satisfies

$$\mathrm{supp}\{\varphi_t\} \subset [0, x_{\max} + T - t]$$

almost surely for all $t \geq 0$.

Proof. For each $x \geq 0$ let

$$\mathcal{T}_x = \{\mu \in F^*; \mathrm{supp}\{\mu\} \subset [0, x]\} \subset F^*.$$

The orthogonal complement of $\mathcal{T}_x$ is the closed subspace given by

$$\mathcal{T}_x^\perp = \{f \in F; f(s) = 0 \text{ for all } s \in [0, x]\} \subset F,$$

and since $\mathcal{T}_x$ is closed, we have $\mathcal{T}_x^{\perp\perp} = \mathcal{T}_x$. Now we show that the strong random operator $Y_{t,T}$ takes $\mathcal{T}_{x+T-t}^\perp$ into $\mathcal{T}_x^\perp$ for each $x \geq 0$. Recall that the process $\{Y_{t,s}\}_{s\geq t}$ satisfies the integral equation

$$Y_{t,s} = S_{t-s} + \int_t^s S_{s-u}\nabla\sigma_u Y_{t,u} \cdot dW_u$$

for $s \geq t$, and note that the random operator $\nabla\sigma(t, \tilde{P}_t)$ takes $\mathcal{T}_x^\perp$ into $\mathcal{L}_{\mathrm{HS}}(G, \mathcal{T}_x^\perp)$ for each $x \geq 0$. Fix $t \in [0, T]$ and consider the Picard iteration scheme $Y_{t,s}^{(0)} = S_{t-s}$ and

$$Y_{t,s}^{(n+1)} = S_{t-s} + \int_t^s S_{s-u}\nabla\sigma_u Y_{t,u}^{(n)} \cdot dW_u.$$

Clearly for all $s \geq t$ the operator $Y_{t,s}^{(0)}$ takes $\mathcal{T}_{x+s-t}^\perp$ into $\mathcal{T}_x^\perp$ for each $x \geq 0$.

Now assuming that $Y_{t,s}^{(n)}$ takes $\mathcal{T}_{x+s-t}^{\perp}$ into $\mathcal{T}_x^{\perp}$, we have that the product $S_{s-u}\nabla\sigma_u Y_{t,u}$ takes $\mathcal{T}_{x+s-t}^{\perp}$ into $\mathcal{L}_{\mathrm{HS}}(G,\mathcal{T}_x^{\perp})$ for all $u \in [t,s]$. And hence the claim is proven by induction and the convergence of the Picard scheme. We note that random vector $\nabla\tilde{g}(\tilde{P}_t)$ is valued in $\mathcal{T}_{x_{\max}}$. Hence for every $f \in \mathcal{T}_{x_{\max}+T-t}^{\perp}$ we have

$$\mathbb{E}\{\langle\nabla\tilde{g}(\tilde{P}_T), Y_{t,T}f\rangle_F|\mathcal{F}_t\} = 0$$

which completes the proof. □

This theorem implies that hedging strategies for this class of contingent claims have the property that the support of the portfolio at almost all times is confined to an interval. Moreover, the right end point of this interval is given by the longest maturity of the bonds underlying the claim, confirming our intuition about maturity specific risk.

Notes & Complements

The fact that the long rate, if it exists, can never fall was first observed by Dybvig, Ingersoll, and Ross [52] in the setting of discretely compounded interest rates. The proof presented in this chapter is modeled after the very succint proof found in the short paper of Hubalek, Klein, and Teichmann [82].

Our discussion of the martingale measures for the HJM model driven by an infinite dimensional Wiener process is patterned after Filipović's PhD. In particular, as far as we know, the example of a concrete state space $F = H_w$ studied in this chapter was introduced by this author. Filipović starts from the evolution form of the HJM dynamical equation, unlike Kusuoka [96] who works directly from the hyperbolic stochastic partial differential equation and the drift condition to show that any solution will force the discounted prices of the zero coupons to be martingales. These works focus on pricing, and as such, they worry about the absence of arbitrage and consequently, about the drift condition.

We have considered exclusively models driven by Wiener processes, and have completely ignored the possibility of jumps in the interest rate dynamics. This choice might not be entirely realistic but allowed us to focus on HJM models driven by infinite dimensional Wiener processes. Of course, interest rate models with jumps have been proposed in the literature; see for instance the paper of Shirakawa [124]. For the simplest case where

$$df_t = (Af_t + \alpha_t)dt + \sigma_t dW_t + \eta_t(dN_t - \lambda_t dt),$$

the forward rates are driven by a Wiener process and a d-dimensional Poisson process $\{N_t\}_{t\geq 0}$ with time-varying intensity $\{\lambda_t\}_{t\geq 0}$, the drift condition becomes

$$\alpha_t(x) = F_{\mathrm{HJM}}(\sigma_t)(x) + \sum_{i=1}^{d} \lambda_t^{(i)}\eta_t^{(i)}(x)\left(1 - e^{-\int_0^x \eta_t^{(i)}(s)ds}\right).$$

The drift condition for HJM-type models with marked point processes has been studied by Björk, Kabanov, and Runggaldier [13]. Eberlein, Jacod, and Raible considered term structure models driven by a Lévy process, and find that such term structure models are complete if and essentially only if the driving Lévy process is one-dimensional.

The existence of finite-dimensional realizations of HJM interest rate models and related geometrical questions have been extensively studied by Björk, and his coauthors Christensen, Gombani, Landén, and Svensson [15], [14], [17], [12], and [11]. Björk has written very accessible surveys of this literature [9] and [10], with many examples worked out in detail. The state space in the above cited works is typically a Hilbert subspace of the space of real-analytic functions. Such spaces are extremely small, and in fact too small to include the important example of the CIR model. Filipović and Teichmann in [64], [63], and [65] have generalized these results to much larger spaces of C^∞ functions. Filipović [61] and Zabczyk [137] have studied the related issue of existence of invariant manifolds for the mild solutions of SPDE.

Our presentation of generalized bond portfolios is modeled after the work of Björk, Di Masi, Kabanov, and Runggaldier [16], and De Donno and Pratelli [48]. Carmona and Tehranchi consider in [36] the hedging problem for a interest rate contingent claim with portfolios valued in the dual of a Hilbert space and quantified the notion of maturity specific risk. Musiela and Goldys [106] studied a similar problem by finding conditions under which the solution to the associated infinite dimensional Kolmogorov equation is differentiable.

Ekeland and Taflin [53] considered the related problem of utility maximization in an approximately complete bond market and found conditions under which an optimal strategy exists. Ringer and Tehranchi [116] provide a construction for the optimal strategy in the case of a certain Gaussian random field model. De Donno [47] and Taflin [128] discussed in detail the set of claims which can be replicated with generalized bond portfolios.

7

Specific Models

In this chapter we study, in the context of infinite-dimensional stochastic analysis, the features of some interest rate models proposed in the literature. We study the following issues in this order: the mean-reverting behavior of Gauss–Markov HJM models, the implications of a certain parabolic stochastic partial differential equation borrowed from random string theory and used as a model for the term structure, and finally, the market model of Brace–Garatek–Musiela for the LIBOR rates.

7.1 Markovian HJM Models

We say that a model is Markovian if the randomness in the drift and volatility coefficients comes only through the randomness of the forward curve itself. More precisely, the model is Markovian if there exist deterministic functions $\bar{\alpha}$ and $\bar{\sigma}$ on $\mathbb{R}_+ \times F$ such that:

$$\alpha(t, \omega, f) = \bar{\alpha}(t, f), \qquad \text{and} \qquad \sigma(t, \omega, f) = \bar{\sigma}(t, f). \tag{7.1}$$

We already encountered Markovian models many times in the previous chapters. In what follows, we shall not use the bar notation to avoid overloading the notation. The stochastic differential equation becomes:

$$df_t = (Af_t + \alpha(t, f_t))dt + \sigma(t, f_t)dW_t,$$

its evolution form reads:

$$f_t = S_t f_0 + \int_0^t S_{t-s}\alpha(s, f_s)\, ds + \int_0^t S_{t-s}\sigma(s, f_s)dW_s,$$

and the usual Lipschitz conditions guarantee existence and uniqueness of the solution of this evolution form.

Obviously, the terminology is justified by the fact that, when the drift and volatility coefficients are of this form, then the solution process $\{f_t;\, t \geq 0\}$ is a Markov process in the state space F.

7.1.1 Gaussian Markov Models

A further restriction often put on the drift and volatility coefficients is to assume that they are both deterministic. One can think of such a model as a Markovian model where the coefficients do not depend upon the forward rate or the calendar time. In this case, the stochastic differential equation looks like:

$$df_t = (Af_t + \alpha)dt + \sigma dW_t \tag{7.2}$$

and its evolution form reads:

$$f_t = S_t f_0 + \int_0^t S_{t-s}\alpha\, ds + \int_0^t S_{t-s}\sigma dW_s \tag{7.3}$$

but, even though these equations look quite like the equations (6.5) and (6.6) of the generalized HJM models in the Musiela's notation, they are quite different because the coefficients are now *deterministic*. Because of this non-randomness, and because the initial condition is as usual assumed to be independent of the Wiener process W, (7.3) is not an equation since its right-hand side determines completely the forward rate process. Notice that this process is Gaussian whenever the initial condition f_0 is Gaussian or deterministic. This justifies the terminology used in the title of this subsection. In fact the evolution form (7.3) shows that $\{f_t;\, t \geq 0\}$ is an Ornstein–Uhlenbeck process in the space F.

One of the niceties of the Markov processes is to provide convenient tools for the analysis of the long-time behavior of the sample paths of the process. We shall consider this problem only in the case of deterministic coefficients, for in this case, the evolution Eq. (7.3) gives:

$$\mu_t = (S_t\mu_0) * (\tilde{\alpha}_t + \gamma_t) \tag{7.4}$$

where μ_t denotes the distribution of f_t while

$$\tilde{\alpha}_t = \int_0^t S_{t-s}\alpha\, ds$$

denotes a nonrandom element of F, and γ_t denotes the distribution of the stochastic integral

$$\int_0^t S_{t-s}\sigma dW_s$$

appearing in the right-hand side of (7.3). The convolution of probability distribution is here because of the independence of the initial condition f_0 and the Wiener process. We shall try to understand the ergodic behavior of the solutions of the evolution form of the generalized HJM equation by controlling the limit for large times of the distribution μ_t. We shall do this by addressing separately the following three problems:

- Does the limit $\tilde{\alpha} = \lim_{t\to\infty} \tilde{\alpha}_t$ exist as an element of F?
- Does the limit $\gamma = \lim_{t\to\infty} \gamma_t$ exist as a probability measure on F?
- Does the limit $\nu = \lim_{t\to\infty} S_t\mu_0$ exist as a probability measure on F?

If we can answer these three questions in the affirmative, we have:

$$\mu_\infty = \nu * (\tilde{\alpha} + \gamma)$$

which gives an invariant measure obtained as a mixture of shifts by elements of the support of ν, of the Gaussian measure γ shifted to have mean $\tilde{\alpha}$.

7.1.2 Assumptions on the State Space

We now list our assumptions on the state space F.

Assumption 7.1. 1. The space F is a separable Hilbert space and the elements of F are continuous, real-valued functions on $\mathbb{R}_+$. The evaluation functional $\delta_x \in F^*$ are continuous linear functionals on F.
2. For every $f \in F$ the limit $f(\infty) = \lim_{x\to\infty} f(x)$ exists, and the functional $\delta_\infty : f \hookrightarrow f(\infty)$ is an element of F^*. Let

$$F^0 = \{f \in F; f(\infty) = 0\}$$

be the closed subspace of F orthogonal to the functional δ_∞.
3. The semigroup $\{S_t\}_{t\geq 0}$ is strongly continuous on F and each operator S_t is a contraction on F^0. Moreover, there exists constants $\beta > 0$ and $M \geq 1$ such that

$$\|S_t f\| \leq M e^{-\beta t} \|f\|$$

for every $f \in F^0$.
4. The binary operator $\star$ defined by

$$(f \star g)(x) = f(x) \int_0^x g(s)ds$$

is such that for all $f, g \in F^0$ the following bound holds:

$$\|f \star g\|_F \leq C \|f\|_F \|g\|_F$$

for some constant $C > 0$.

Assumption 7.1.2 ensures that our HJM models on F will generate forward rate curves such that the long rate exists. Notice that since the drift α is assumed to be an element of F, in particular $\alpha(\infty) = \lim_{x\to\infty}\langle \delta_x^* \sigma, I_x^* \sigma + \lambda\rangle$ exists. Thus we must restrict the range of the operator σ to be contained in F^0. Assumption 7.1.4 says that this is the only restriction we need on the choice of volatility. Indeed, by Proposition 6.1, the map F_{HJM} is Lipschitz continuous from $\mathcal{L}_{\text{HS}}(G, F^0)$ into F. But since for $f, g \in F^0$ we have

$$\left| f(x) \int_0^x g(s)ds \right| \le |f(x)| \|g\| \int_0^x \|S_s\| \|\delta_0\| ds$$
$$\le |f(x)| \|g\| \|\delta_0\| \frac{M}{\beta}$$

which converges toward 0 as $x \to \infty$, we have that the vector α is actually in the subspace F^0; that is $F_{\mathrm{HJM}}(\sigma)(\infty) = 0$ for all $\sigma \in \mathcal{L}_{\mathrm{HS}}(G, F^0)$.

Remark 7.1. The space H_w studied in depth in Sect. 6.3.3 satisfies Assumption 7.1 if the weight function grows fast enough. In particular, if the weight w satisfies the bound

$$\frac{1}{2} \inf_{x \ge 0} \frac{w'(x)}{w(x)} = \beta > 0$$

then H_w satisfies Assumption 7.1.3. Indeed, for $f \in H_w^0$ we have

$$\|S_t f\|^2 = f(t)^2 + \int_0^\infty w(x) f'(x+t)^2 dx$$
$$\le \|f\|^2 \left(\int_t^\infty w(x)^{-1} dx + \sup_{x \ge 0} \frac{w(x)}{w(x+t)} \right)$$
$$\le \|f\|^2 e^{-2\beta t} \left((2\beta w(0))^{-1} + 1 \right)$$

as desired.

7.1.3 Invariant Measures for Gauss–Markov HJM Models

We now come to the main result of this section:

Theorem 7.1. *Let $\lambda \in G$ and $\sigma \in \mathcal{L}_{\mathrm{HS}}(G, F^0)$ be deterministic constants. The HJM model given by the equation*

$$f_t = S_t f_0 + \int_0^t S_{t-s}(F_{\mathrm{HJM}}(\sigma) + \sigma\lambda) ds + \int_0^t S_{t-s} \sigma dW_s \tag{7.5}$$

has a family of invariant measures $\{\mu^\nu\}_\nu$ on the space F. For each fixed ν, the random variables $f(x) = \delta_x(f)$ on $(F, \mathcal{B}_F, \mu^\nu)$ have the following properties:

1. *The distribution of the long rate $f(\infty)$ is ν.*
2. *The conditional distribution of the forward rate $f(x)$ given the long rate is Gaussian, for every $x \ge 0$.*
3. *If $\int_{-\infty}^\infty |c| \ \nu(dc) < +\infty$ and if $\lambda = 0$ then the expected value of the short rate $f(0)$ is greater than or equal to the expected value of any other forward rate.*
4. *If $\int_{-\infty}^\infty c^2 \ \nu(dc) < +\infty$ then the variance of the short rate is greater than or equal to the variance of any other forward rate.*

Proof. The space F decomposes into the direct sum $F = F^0 \oplus \mathbb{R}\mathbf{1}$. Notice that for every realization of the initial term structure, $S_t f_0 = \mathbf{1} f_0(\infty) + S_t \mathrm{Proj}_{F^0} f_0$ converges strongly to the element $\mathbf{1} f_0(\infty) \in F$. Hence, the measure $S_t \mu_0$ converges weakly to the measure ν supported on the subspace of F spanned by the constant functions. Abusing notation slightly, we can identify ν as the measure on $\mathbb{R}$ given by the marginal distribution of the initial long rate. Now since

$$\left\| \int_0^t S_{t-s} \alpha ds \right\| \leq \frac{M\|\alpha\|}{\beta},$$

the Bochner integral

$$\tilde{\alpha}_t = \int_0^t S_{t-s} \alpha ds$$

converges strongly to an element $\alpha \in F^0$. Similarly, since

$$\int_0^t \|S_{t-s}\sigma\|^2 ds \leq \frac{M\|\sigma\|^2}{2\beta}$$

the law γ_t of the stochastic integral $\int_0^t S_{t-s}\sigma dW_s$ converges weakly to the mean-zero Gaussian measure γ supported on the subspace F^0 with covariance $\int_0^\infty S_s \sigma \sigma^* S_s^* ds$.

By the continuity of the evaluation functionals δ_x we have

$$\begin{aligned}
\mathbb{E}\{f_t(x)\} - \mathbb{E}\{f_0(t+x)\} &= \delta_x \int_0^t S_{t-s}(F_{\mathrm{HJM}}(\sigma) + \sigma\lambda) ds \\
&= \int_0^t \delta_{t-s+x}(F_{\mathrm{HJM}}\sigma) + \sigma\lambda) ds \\
&= \int_x^{t+x} \langle \sigma^*\delta_s, \sigma^* I_s + \lambda \rangle_G ds \\
&= \frac{1}{2}\left(\|\sigma^* I_{x+t} + \lambda\|_G^2 - \|\sigma^* I_x + \lambda\|_G^2 \right).
\end{aligned}$$

Noting that the functional δ_{t+x} converges weakly in F^* to δ_∞ and the functional I_{t+x} converges weakly in F^{0*} to a functional $I_\infty \in F^{0*}$ we have

$$\begin{aligned}
\mathbb{E}_{\mu^\nu}\{f(x)\} &= \lim_{t\to\infty} \mathbb{E}\{f_t(x)\} \\
&= \mathbb{E}_{\mu^\nu}\{f(\infty)\} + \frac{1}{2}\left(\|\sigma^* I_\infty + \lambda\|_G^2 - \|\sigma^* I_x + \lambda\|_G^2 \right).
\end{aligned}$$

In the case that $\lambda = 0$, it follows that

$$\begin{aligned}
\mathbb{E}_{\mu^\nu}\{f(x)\} &= \mathbb{E}_{\mu^\nu}\{f(\infty)\} + \frac{1}{2}\left(\|\sigma^* I_\infty\|_G^2 - \|\sigma^* I_x\|_G^2 \right) \\
&\leq \mathbb{E}_{\mu^\nu}\{f(\infty)\} + \frac{1}{2}\|\sigma^* I_\infty\|_G^2 = \mathbb{E}_{\mu^\nu}\{f(0)\}.
\end{aligned}$$

Finally, the variance of $f(x)$ is given by

$$\mathrm{Var}_{\mu^\nu}\{f(x)\} = \mathrm{Var}_{\mu^\nu}\{f(\infty)\} + \int_x^\infty \|\sigma^* \delta_t\|_G^2 dt$$

which is decreasing in $x \geq 0$. □

Remark 7.2. We used freely the results on weak convergence of Gaussian measures which were reviewed in the Notes & Complements section of Chap. 3.

Remark 7.3. We note that setting $\lambda = 0$ in the above theorem is equivalent to considering the risk-neutral HJM dynamics given by

$$f_t = S_t f_0 + \int_0^t S_{t-s} F_{\mathrm{HJM}}(\sigma)\, ds + \int_0^t S_{t-s}\sigma\, d\tilde{W}_s \tag{7.6}$$

where the process $\tilde{W}_t = W_t + \int_0^t \lambda_s ds$ is a cylindrical G-valued Wiener process under the risk-neutral measure $\mathbb{Q}$ with Radon–Nykodym derivative

$$\frac{d\mathbb{Q}}{d\mathbb{P}}\Big|_{\mathcal{F}_t} = \exp\left(-\frac{1}{2}\int_0^t \|\lambda_s\|_G^2\, ds + \int_0^t \lambda_s d\tilde{W}_s\right).$$

7.1.4 Non-Uniqueness of the Invariant Measure

Mathematically, there is no surprise that there are an infinite number of invariant measures for the HJM equation. Indeed, by the discussion of the finite dimensional Ornstein–Uhlenbeck process in Chap. 4, we should expect many invariant measures if the drift operator A has a nontrivial kernel. In our case the restriction of A to the differentiable elements of F is just the derivative d/dx. And since the derivative of a constant function is the zero function, the kernel of A is nontrivial.

On an economic level, it might seem strange that the HJM dynamics are not ergodic and that the effects of the initial data are not forgotten over time. In a practical sense, however, the HJM model does admit a unique invariant measure. Indeed, it is meaningless to consider initial forward curves with differing long rates since, within the context of a given HJM model, the long rate is constant. In other words, the value $c = f_0(\infty)$ can be considered a model parameter, elevated to the status of the functions λ and σ. Given the three parameters λ, σ, and c, the unique invariant measure for the HJM model is the measure μ^ν from the theorem, where ν is the point mass concentrated at c.

For the sake of comparison, recall that the popular Vasicek and CIR short rate models are ergodic. One may wonder why the long term behavior is so different for HJM models. The answer is that in both cases the density of the invariant measure (Gaussian for the Vasicek model and non-central χ^2 for the CIR model) depends on the model parameters. Since the parameters are chosen in part to fit the initial term structure, the effects of the initial data are really never forgotten after all.

7.1.5 Asymptotic Behavior

For large times t (ergodic theorem) we expect f_t to look like an element f of F drawn at random according to the distribution μ. Such a random sample f is obtained as follows:

- Choose a level for $c = f(\infty)$ at random according to ν.
- Shift $\tilde{\alpha}$ to give it this value c at the limit as $x \to \infty$.
- Perturb this candidate for f by a random element of F generated according to the Gaussian distribution γ.

If we choose a generalized HJM model for the historical dynamics of the forward curve, then the diagonalization of the covariance operator of γ should fit the empirical results of the PCA:

- The eigenvalues of this covariance operator should decay at the same rate as the (empirical) proportions of the variance explained by the principal components.
- The eigenfunctions corresponding to the largest eigenvalues should look like the main loadings of the PCA.

7.1.6 The Short Rate is a Maximum on Average

Consider a forward rate process $\{f_t\}_{t\geq 0}$ arising from a generalized HJM model. In this section we study the invariant measure for the risk-neutral dynamics when the forward rates are an F-valued time-homogeneous Markov process. In the previous section, we observed a curious property of invariant measures in the context of Gauss–Markov HJM models: the short rate is a maximum on average for the forward rate curve. This property is very general, and the proof requires very little structure on the forward rate dynamics.

We assume that the discounted bond prices are martingales for the risk neutral measure $\mathbb{Q}$. In particular, the following relationship holds:

$$\exp\left(-\int_0^x f_t(s)ds\right) = \mathbb{E}\left\{ \exp\left(-\int_t^{x+t} r_s ds\right) |\mathcal{F}_t\right\}$$

where $r_t = f_t(0)$. The precise claim is:

Proposition 7.1. *Let μ be an invariant measure for the Markov process $\{f_t\}_{t\geq 0}$. Let*

$$\bar{f}(x) = \int_F f(x)\mu(df)$$

be the average forward rate with time to maturity x for this measure, and let us assume that $\bar{f}$ is continuous. Then the average short rate is a maximum as:

$$\bar{f}(0) \geq \bar{f}(x)$$

for all $x \geq 0$.

Proof. Fix $\epsilon > 0$. Then,

$$\begin{aligned}\exp\left(-\int_0^{x+\epsilon} f_0(s)ds\right) &= \mathbb{E}\left\{\exp\left(-\int_0^{x+\epsilon} r_s ds\right)|\mathcal{F}_0\right\}\\ &= \mathbb{E}\left\{\exp\left(-\int_0^{\epsilon} r_s ds\right)\mathbb{E}\left\{\exp\left(-\int_\epsilon^{x+\epsilon} r_s ds\right)|\mathcal{F}_\epsilon\right\}|\mathcal{F}_0\right\}\\ &= \mathbb{E}\left\{\exp\left(-\int_0^{\epsilon} r_s ds - \int_0^x f_\epsilon(s)ds\right)|\mathcal{F}_0\right\}\\ &\geq \exp\left(-\mathbb{E}\left\{\int_0^{\epsilon} r_s ds|\mathcal{F}_0\right\} - \mathbb{E}\left\{\int_0^x f_\epsilon(s)ds|\mathcal{F}_0\right\}\right)\end{aligned}$$

by Jensen's inequality. Rearranging the above inequality, we have

$$\mathbb{E}\left\{\int_0^{\epsilon} r_s ds|\mathcal{F}_0\right\} \geq \int_0^{x+\epsilon} f_0(s)ds - \mathbb{E}\left\{\int_0^x f_\epsilon(s)ds|\mathcal{F}_0\right\}.$$

Now, integrating both sides with respect to the invariant measure μ and using Fubini's theorem:

$$\epsilon\bar{f}(0) \geq \int_x^{x+\epsilon} \bar{f}(s)ds.$$

Dividing by ϵ and then letting $\epsilon \searrow 0$ proves the claim. □

The average forward curve $\bar{f}$ considered in this subsection can be interpreted as a steady-state forward curve obtained when the system is given enough time to relax to equilibrium. From the above proposition, one might suspect that the average forward curve $\bar{f}$ is a decreasing function, but this is not always the case. Nevertheless, if we limit ourselves to HJM models with the economically reasonable property that the infinitesimal increments of the forward rates are positively correlated, then as we are about to prove, the average forward curve $\bar{f}$ is indeed decreasing.

Let the forward rate process $\{f_t\}_{t\geq 0}$ be an F-valued time-homogeneous Markov process formally solving the following HJM equation:

$$df_t = (A\ f_t + F_{\mathrm{HJM}} \circ \sigma(f_t))dt + \sigma(f_t)dW_t$$

where $\{W_t\}_{t\geq 0}$ is a Wiener process defined cylindrically on a separable Hilbert space G, the operator A is the generator of the semigroup $\{S_t\}_{t\geq 0}$ of left shifts and is such that $Ag = g'$ whenever g is differentiable, and the volatility function $\sigma : F \to \mathcal{L}_{\mathrm{HS}}(G, F)$ is such that measure μ on F is invariant for the dynamics of $\{f_t\}_{t\geq 0}$. Furthermore, we assume that σ is bounded for the sake of simplicity.

Proposition 7.2. *Under the above assumptions, if we denote as usual by* $\delta_x \in F^*$ *the evaluation functional* $\delta_x(g) = g(x)$ *and if we assume that*

$$\langle\sigma(f)^*\delta_x, \sigma(f)^*\delta_y\rangle \geq 0$$

for all $f \in F$ *and* $x, y \in \mathbb{R}_+$, *then the average forward curve* $\bar{f}$ *is decreasing.*

Proof. The HJM equation can be rewritten as

$$f_t(x) = f_0(t+x)+\int_0^t \langle \sigma(f_u)^*\delta_{t-u+x}, \sigma(f_u)^* I_{t-u+x}\rangle du+\int_0^t \sigma(f_u)^*\delta_{t-u+x}dW_u$$

where $I_s \in F$ is the definite integral functional defined by $I_x(g) = \int_0^x g(u)du$. Supposing the law of f_0 is the invariant measure μ and taking expectations of both sides, we have

$$\bar{f}(x) = \bar{f}(x+t) + \mathbb{E}\left\{\int_0^t \langle \sigma(f_0)^*\delta_{u+x}, \sigma(f_0)^* I_{u+x}\rangle du\right\}.$$

Since $\langle \sigma(f)^*\delta_s, \sigma(f)^* I_s\rangle = \int_0^s \langle \sigma(f)^*\delta_s, \sigma(f)^*\delta_t\rangle dt \geq 0$ by assumption, the result follows. □

The above proposition might seem surprising. Yield curves, after all, are much more likely to be upward sloping than downward sloping. In fact, economists say that the yield curve is *inverted* when it is downward sloping, a term which suggests that such events are considered deviations from the norm. However, we have considered the dynamics of the forward rates under the risk-neutral measure $\mathbb{Q}$, not the historical measure $\mathbb{P}$. It follows then that the risk-neutral measure assigns much larger weight to the events when the yield curves are downward sloping than the historical measure.

7.2 SPDE's and Term Structure Models

Rewriting a generalized HJM model as a stochastic partial differential equation has been done in two very different ways.

The first approach is to accept the drift condition as given, and work directly with the hyperbolic SPDE. In a clever tour de force, Kusuoka showed in [96] (essentially with his bare hands) that, for any solution in the sense of Schwartz distributions of this SPDE, the discounted prices of the zero-coupon bonds are necessarily martingales, that essentially never vanish, and that if they do, they remain zero from the time they vanish on.

The second approach is motivated by risk control. In order to run Monte-Carlo simulations to evaluate the risk of fixed income portfolios, a good understanding of the dynamics of the term structure under the historical probability measure $\mathbb{P}$ is necessary. So instead of trying to enforce the HJM drift condition, the modeling emphasis is on replicating the statics, such as the PCA, of the forward rate curves found in the market.

We work under the assumption that the domain of the forward rate function $x \hookrightarrow f_t(x)$, in the Musiela notation, is the bounded interval $[0, x_{\max}]$, as opposed to the half line $\mathbb{R}_+$. One motivation for working on a bounded interval is practical: there just are not very many Treasuries with maturities greater than thirty years! We may and do choose the units of time such that $x_{\max} = 1$ without loss of generality. As usual, the left-hand endpoint $f_t(0) = r_t$ is the short rate and the right-hand endpoint $f_t(1) = \ell_t$ is the long rate. Let $s_t = \ell_t - r_t$ be the spread. This use of the term spread should not be confused with the difference of the bid and ask price of an asset. The forward rate is then decomposed into

$$f_t(x) = r_t + s_t[y(x) + g_t(x)]$$

where $x \hookrightarrow y(x)$ is a deterministic function with $y(0) = 0$ and $y(1) = 1$ which is used to capture the "average" shape of the forward rate curve. An example of such a function y which seems to agree with market data is $y(x) = \sqrt{x}$. The random function $x \hookrightarrow g_t(x)$ is thought of as the deformations of this average profile. Notice that by construction, the boundary conditions $g_t(0) = g_t(1) = 0$ are satisfied for all $(t, \omega) \in \mathbb{R}_+ \times \Omega$. As time progresses, the graphs of the functions $x \hookrightarrow g_t(x)$ resemble a random vibrating string with fixed endpoints.

7.2.1 The Deformation Process

We work in the context of an HJM model with the forward rate dynamics given formally by the equation

$$df_t = \left(\frac{\partial}{\partial x} f_t + \alpha_t\right) dt + \sigma_t dW_t$$

where $\{W_t\}_{t\geq 0}$ is a Wiener process defined cylindrically on a separable Hilbert space G, the drift process $\{\alpha_t\}$ takes values in some Hilbert space F, and the volatility process $\{\sigma_t\}_{t\geq 0}$ takes values in $\mathcal{L}_{\mathrm{HS}}(G, F)$. The drift and volatility are related by the condition $\alpha_t = F_{\mathrm{HJM}}(\sigma_t) + \sigma_t \lambda_t$ for some G-valued process $\{\lambda_t\}_{t\geq 0}$. The short rate r_t then formally satisfies the equation

$$dr_t = \left(\frac{\partial}{\partial x} f_t(0) + \alpha_t(0)\right) dt + \sigma_t(0) dW_t$$

where $\sigma_t(0) = \sigma_t^* \delta_0$. Similarly, the long rate ℓ_t satisfies

$$d\ell_t = \left(\frac{\partial}{\partial x} f_t(1) + \alpha_t(1)\right) dt + \sigma_t(1) dW_t.$$

Now applying Itô's formula to the deformation

$$g_t(x) = \frac{f_t(x) - r_t}{s_t} + y(x),$$

where $s_t = \ell_t - r_t$ is the spread, we see that g_t formally satisfies an equation of the form

$$dg_t = \left(\frac{\partial}{\partial x} g_t + m_t\right) dt + n_t dW_t$$

where the volatility is given by

$$n_t = s_t^{-1}(\sigma_t - \mathbf{1} \otimes \sigma_t(0)) - s_t^{-2}(f_t - r_t) \otimes (\sigma_t(1) - \sigma_t(0))$$

and the drift m_t is given by an even more complicated formula.

7.2.2 A Model of the Deformation Process

Breaking from the no-arbitrage framework, Cont proposes in [41] to model the deformation process $\{g_t(x)\}_{t\geq 0, x\in[0,1]}$ independently of the short and long rates. As usual, the two-parameter random field is interpreted as a one-parameter stochastic process $\{g_t\}_{t\geq 0}$ where $g_t = g_t(\cdot)$ takes values in a function space H. It turns out that a convenient state space is given by the Hilbert space $H = L^2([0,1], e^{2x/k}dx)$. The process is assumed to be the mild solution of the following SPDE:

$$dg_t = A^{(\kappa)} g_t dt + \sigma_0 dW_t \tag{7.7}$$

where $\kappa > 0$ and $\sigma_0 > 0$ are scalar constants and the partial differential operator $A^{(\kappa)}$ is given by

$$A^{(\kappa)} = \frac{\partial}{\partial x} + \frac{\kappa}{2}\frac{\partial^2}{\partial x^2}. \tag{7.8}$$

The Wiener process $\{W_t\}_{t\geq 0}$ is defined cylindrically on H, and assumed to be independent of the diffusion $\{(r_t, s_t)\}_{t\geq 0}$.

By using Eq. (7.7) as the model for the deformation process, we have modified the no-arbitrage dynamics in three significant ways:

- We have replaced the Hilbert–Schmidt operator n_t with the constant multiple of the identity $\sigma_0 I$.
- We have replaced the operator $A = \partial/\partial x$ by the operator $A^{(\kappa)}$.
- We have also completely ignored the term m which arises from the HJM no-arbitrage drift condition.

Even though we are convinced that our abuse of notation will not create confusion, it is important to notice that we are working in a state space H that does not satisfy the Assumption 6.1. In particular, pointwise evaluation is not continuous on H.

To lighten notation, fix $\kappa > 0$ and let $A = A^{(\kappa)}$ be the self-adjoint extension of the partial differential operator $A^{(\kappa)}$ defined in formula (7.8) with Dirichlet boundary conditions. Notice that imposing these boundary conditions will guarantee that $g(1) = g(0) = 0$ whenever $g \in \mathcal{D}(A)$. Now to get a handle on the effects of these modifications, we rewrite Eq. (7.7) in the evolutionary form as

$$g_t = S_t g_0 + \sigma_0 \int_0^t S_{t-s} dW_s \tag{7.9}$$

where $\{S_t\}_{t\geq 0}$ is the semigroup generated by A. Although the semigroup $\{S_t\}_{t\geq 0}$ is strongly continuous on H, Eq. (7.9) is fundamentally different than the stochastic evolution equations we have faced before since the volatility operator $\sigma_0 I$ is *not* Hilbert–Schmidt. Indeed, it is not clear a-priori that the stochastic integral on the right-hand side of the equation is well-defined. Fortunately, the integral is well-defined as we will see, thanks to the regularizing effects of the operator A. Indeed, replacing the operator $\partial/\partial x$ with the operator A by adding the *viscosity term* given by the Laplacian does wonders. In particular, it turns a hyperbolic SPDE into a regularizing parabolic SPDE. This new SPDE is of the Ornstein–Uhlenbeck type.

Finally, it is important to point out that by ignoring the drift term m this model is not an HJM model: there does not exist an equivalent measure such that all of the discounted bond prices are simultaneously martingales. The model is therefore not appropriate for asset pricing. The justification for such a radical departure from the HJM framework is that this model exhibits many of the stylized empirical features of the forward rates, and hence is appropriate for risk management. Furthermore, the model is quite parsimonious. There are only two parameters, σ_0 and κ, to be estimated from market data.

7.2.3 Analysis of the SPDE

Before we analyze the SDE of Eq. (7.7) or its evolutionary form (7.9), we identify the domain of the unbounded operator A with boundary conditions. It is the subspace of H consisting of functions on $[0, 1]$ vanishing at 0 and 1, which are absolutely continuous together with their first derivatives and such that the second derivatives belong to H. Because of the presence of the exponential weight in the measure defining H as an L^2-space, plain integration by parts shows that the operator A is self-adjoint on its domain. Its spectrum is discrete and strictly negative. The n-th eigenvalue $-\lambda_n$ is given by

$$\lambda_n = \frac{1}{2\kappa}(1 + n^2\pi^2\kappa^2)$$

with corresponding normalized eigenvector

$$e_n(x) = \sqrt{2}\sin(n\pi x)e^{-x/\kappa}.$$

The eigenvectors form an orthonormal basis $\{e_n\}_n$ of H. Notice that the graphs of the first few eigenfunctions share an uncanny resemblance to the first few factors found in the principal component analysis of US interest rate data, as reported in Chap. 1. Of course, this is no coincidence: The operator A and the Hilbert space H were chosen with these stylized empirical facts in mind.

The semigroup $\{S_t\}_{t\geq 0}$ generated by A is analytic. Furthermore, for $t > 0$, the operator S_t can be decomposed into the norm-converging sum

$$S_t = \sum_{n=1}^{\infty} e^{-\lambda_n t} e_n \otimes e_n$$

and hence we have the equality

$$\|S_t\|_{\mathcal{L}_{\mathrm{HS}}(H)} = \left(\sum_{n=1}^{\infty} e^{-2\lambda_n t} \right)^{1/2}.$$

And because we have the bound

$$\begin{aligned}\int_0^t \|S_s\|^2_{\mathcal{L}_{\mathrm{HS}}(H)} ds &= \int_0^t \sum_{n=1}^{\infty} e^{-2t\lambda_n} dt \\ &\leq \sum_{n=1}^{\infty} \frac{\kappa}{1+n^2\pi^2\kappa^2} < +\infty\end{aligned}$$

the stochastic integral $\int_0^t S_{t-s} dW_s$ makes sense as promised. The stochastic integral defines a continuous process in fact, but we shall see that this process is *not* a semi-martingale.

Let $g_t^{(n)} = \int_0^1 g_t(x) e_n(x) e^{2x/\kappa} dx$ be the projection of the deformation g_t onto the n-th eigenvector. The projections $\{g^{(n)}\}_n$ are scalar Ornstein–Uhlenbeck processes given by the SDEs

$$dg_t^{(n)} = -\lambda_n g_t^{(n)} + \sigma_0 dw_t^{(n)}$$

where $w_t^{(n)} = \langle e_n, W_t \rangle$ defines a standard scalar Wiener process. And since the eigenvectors $\{e_n\}_n$ are orthogonal, the Wiener processes $\{w^{(n)}\}_n$ are independent. Hence, the deformation can be decomposed into the $L^2(\Omega; H)$ converging sum

$$g_t = \sum_{n=1}^{\infty} e_n g_t^{(n)}$$

of independent and H-valued random variables given explicitly by

$$g_t^{(n)} = e^{-t\lambda_n} g_0^{(n)} + \sigma_0 \int_0^t e^{-(t-s)\lambda_n} dw_t^{(n)}.$$

The first term in the right-hand side of the above equation can be interpreted as follows: the forward rate curve forgets the contribution of the initial term structure to the n-th eigenmode at an exponential rate with a characteristic time given by λ_n^{-1}. In particular, very singular perturbations to the initial term structure decay away much more quickly than smooth ones.

Note as $t \to \infty$, each of the $g_t^{(n)}$ converges in law to a mean-zero Gaussian with variance

$$\frac{\sigma_0^2}{2\lambda_n} = \frac{\sigma_0^2 \kappa}{1+n^2\pi^2\kappa^2}.$$

These variances, which correspond to the loadings of the principal components, decay quickly to zero. Again, this fact agrees nicely with the PCA of Chap. 1.

7.2.4 Regularity of the Solutions

We now turn to the issue of smoothness of the solutions. Recall that the state space $H = L^2([0,1], e^{2x/k}dx)$ is a space of equivalence classes of measurable functions, and a-priori is does not make sense to talk about $g_t(x)$ for fixed $x \in [0,1]$. Indeed, the evaluation functional δ_x is not continuous on H. This possible lack of smoothness has another annoying practical implication: we know that the boundary conditions $g(0) = g(1) = 0$ are satisfied when $g \in \mathcal{D}(A)$ but not for general $g \in H$.

Fortunately, because of the smoothing properties of the Laplacian, there is a version of the random field $\{g_t(x)\}_{t\geq 0, x\in[0,1]}$ which is continuous in the two parameters. In fact more is true, as we see from the following theorem:

Theorem 7.2. *There exists a version of the random field* $\{g_t(x)\}_{t\geq 0, x\in[0,1]}$ *such that almost surely:*

- *for* $t > 0$*, the map* $x \hookrightarrow g_t(x)$ *is Hölder* $\frac{1}{2} - \epsilon$ *continuous for any* $\epsilon > 0$
- *for* $x \in [0,1]$*, the map* $t \hookrightarrow g_t(x)$ *is Hölder* $\frac{1}{4} - \epsilon$ *continuous for any* $\epsilon > 0$.

For the proof, we need the following estimate:

Lemma 7.1. *For real numbers* $p \geq 0$ *and* $q > 1$*, there is a constant* $C > 0$ *such that we have the bound*

$$\sum_{n=1}^{\infty} n^{-q} \wedge (xn^p) \leq Cx^{\frac{q-1}{p+q}}$$

for all $x \geq 0$.

Proof. The inequality clearly holds for $x \geq 1$ with $C = \sum_{n=1}^{\infty} n^{-q} < \frac{q}{q-1}$. So suppose that $x < 1$. Then we have

$$\begin{aligned}\sum_{n=1}^{\infty} n^{-q} \wedge (xn^p) &= \sum_{n < x^{\frac{-1}{p+q}}} xn^p + \sum_{n \geq x^{\frac{-1}{p+q}}} n^{-q} \\ &\leq \int_0^{\left\lfloor x^{\frac{-1}{p+q}} \right\rfloor + 1} xt^p dt + \int_{\left\lfloor x^{\frac{-1}{p+q}} \right\rfloor}^{\infty} t^{-q} dt \\ &\leq \left(\frac{2^{p+1}}{p+1} + \frac{2^{q-1}}{q-1} \right) x^{\frac{q-1}{p+q}}\end{aligned}$$

where $\lfloor y \rfloor$ denotes the greatest integer smaller than y and we have used the inequality $\frac{1}{2}(\lfloor y \rfloor + 1) \leq y \leq 2\lfloor y \rfloor$ for $y > 1$. □

Proof of Theorem 7.2. We have the decomposition

$$g_t = S_t g_0 + \sigma_0 \int_0^t S_{t-s} dW_s.$$

For every $t > 0$, there is a representative of the equivalence class $S_t g_0 \in H$ given by

$$(S_t g_0)(x) = \sqrt{2} \sum_{n=1}^{\infty} \exp\left(- \frac{(1 + n^2\pi^2\kappa^2)t + 2x}{2\kappa} \right) \sin(n\pi x) g_0^{(n)}.$$

It is easy to see that the function $(t, x) \hookrightarrow S_t g_0(x)$ is infinitely differentiable for $t > 0$. Therefore the smoothness of the field $(t, x) \hookrightarrow g_t(x)$ is governed by the stochastic integral, and so from now on we assume $g_0 = 0$.

The eigenfunctions e_n are bounded and Lipschitz, and so satisfy the bounds

$$|e_n(x) - e_n(y)| \leq 2\sqrt{2} \wedge \left(\frac{1}{\kappa} |x - y| \sqrt{2(1 + n^2\pi^2\kappa^2)} \right).$$

Therefore we have

$$\begin{aligned} \mathbb{E}\{(g_t(x) - g_t(y))^2\} &= \sigma_0^2 \sum_{n=1}^{\infty} (e_n(x) - e_n(y))^2 \frac{(1 - e^{-2\lambda_n t})}{2\lambda_n} \\ &\leq C|x - y| \end{aligned}$$

because of the result of Lemma 7.1 with $p = 0$ and $q = 2$. Since the random field is Gaussian there are constants C_n such that we have the bound

$$\mathbb{E}\{(g_t(x) - g_t(y))^{2n}\} \leq C_n |x - y|^n$$

for all $n \geq 1$. The Hölder continuity $x \hookrightarrow g_t(x)$ follows from Kolmogorov's theorem. Similarly, we have

$$\begin{aligned} \mathbb{E}\{(g_s^{(n)} - g_t^{(n)})^2\} &= \frac{1}{2\lambda_n} (2 - 2e^{-\lambda_n |t-s|} - e^{-\lambda_n s} - e^{-\lambda_n t} + 2e^{-\lambda_n (t+s)}) \\ &= \frac{1}{2\lambda_n} (2(1 - e^{-\lambda_n |t-s|}) - (e^{-\lambda_n s/2} - e^{-\lambda_n t/2})^2) \\ &\leq \lambda_n^{-1} \wedge |t - s| \end{aligned}$$

and so

$$\begin{aligned} \mathbb{E}\{(g_s(x) - g_t(x))^2\} &\leq \sigma_0^2 \sum_{n=1}^{\infty} 2e^{-2x/\kappa} \sin(nx)^2 (\lambda_n^{-1} \wedge |t - s|) \\ &\leq C|t - s|^{1/2} \end{aligned}$$

using again Lemma 7.1 with $p = 0$ and $q = 2$. Again, the result follows from Kolmogorov's theorem. □

The above Hölder regularity is essentially the best possible. In particular, for fixed $t > 0$ and $\omega \in \Omega$, the graph of the function $x \hookrightarrow g_t(x)$ more-or-less resembles the sample path of a Brownian bridge . Recall that a Brownian bridge $\{B_x\}_{x\in[0,1]}$ is a mean-zero Gaussian process satisfying the two end-point constraints $B_0 = B_1 = 0$ and with covariance $\mathbb{E}\{B_x B_y\} = x \wedge y - xy$. A Brownian bridge can be realized as $B_x = w_x - x w_1$ where $\{w_x\}_{x\in[0,1]}$ is a standard scalar Wiener process. The name Brownian bridge derives from the simple observation that the Wiener process $w_x = (1-x)w_0 + x w_1 + B_x$ is a linear combination of its endpoint values plus a "bridge" term. Remember that the standard Wiener process has left-hand endpoint $w_0 = 0$. The Brownian bridge has the series expansion (recall the discussion of Sect. 3.5 of Chap. 3)

$$B_x = \sum_{n=1}^{\infty} \frac{\sin(n\pi x)}{n} \xi_n$$

known as Karhunen–Loeve decomposition, where $\{\xi_n\}_n$ are independent standard normal random variables. Comparing this expression with the decomposition

$$g_t(x) = (Tg_0)(x) + \sigma_0 \sqrt{2\kappa} e^{-x/\kappa} \sum_{n=1}^{\infty} (1 - e^{-2t\lambda_n})^{1/2} \frac{\sin(n\pi x)}{(1+\kappa^2\pi^2 n^2)^{1/2}} \xi_n$$

where

$$\xi_n = \frac{g_t^{(n)} - \mathbb{E}\{g_t^{(n)}\}}{\operatorname{var}(g_t^{(n)})^{1/2}}$$

we see that for fixed $t > 0$, we can write the deformation as

$$g_t(x) = \sigma_0 \sqrt{2/\kappa} \pi^{-1} e^{-x/\kappa} B_x + h(x)$$

where $\{B_x\}_{x\in[0,1]}$ is a standard Brownian bridge and the random function $x \hookrightarrow h(x)$ is twice-differentiable almost surely.

We have seen that for fixed $t > 0$, the random function $x \hookrightarrow g_t(x)$ has the same regularity as a Wiener sample path. On the other hand, for fixed $x \in (0,1)$, the regularity of the random function $t \hookrightarrow g_t(x)$ is far worse. Indeed, it can be shown that the *quartic* variation is nonzero, and thus the process $\{g_t(x)\}_{t\geq 0}$ cannot be a semi-martingale!

7.3 Market Models

7.3.1 The Forward Measure

Fix a filtered probability space $(\Omega, \mathcal{F}, \mathbb{P}; \{\mathcal{F}_t\}_{t\geq 0})$, and let ξ be an $\mathcal{F}_T$-measurable random variable. If ξ corresponds to the payout of a contingent claim with maturity $T > 0$, the no-arbitrage theory tells us that the price at time $t \in [0,T]$ of this claim is by the risk-neutral conditional expectation

$$V_t = \mathbb{E}^{\mathbb{Q}}\{e^{-\int_t^T r_s ds} \xi | \mathcal{F}_t\}$$

where r_t is the spot interest rate and $\mathbb{Q} \sim \mathbb{P}$ is an equivalent measure under which the discounted asset prices are local martingales.

If the payout ξ is conditionally independent of the interest rate, we have that the price factors as

$$\begin{aligned} V_t = \mathbb{E}^{\mathbb{Q}}\{e^{-\int_t^T r_s ds}\xi|\mathcal{F}_t\} &= \mathbb{E}^{\mathbb{Q}}\{e^{-\int_t^T r_s ds}|\mathcal{F}_t\}\mathbb{E}^{\mathbb{Q}}\{\xi|\mathcal{F}_t\} \\ &= P(t,T)\mathbb{E}^{\mathbb{Q}}\{\xi|\mathcal{F}_t\} \end{aligned}$$

where $P(t,T)$ is the price of a bond with maturity T. Such a factored representation is convenient since the price $P(t,T)$ is quoted in the market, and so the practitioner can devote all his or her attention to computing the conditional expectation $\mathbb{E}^{\mathbb{Q}}\{\xi|\mathcal{F}_t\}$.

On the other hand, if the payout ξ is contingent on the interest rates at time T, the discounting factor $\exp(-\int_t^T r_s ds)$ can be a nuisance since it depends on the whole history of the short rate process. In particular, the payout ξ and the discounting factor $\exp(-\int_t^T r_s ds)$ cannot be modeled independently.

One way to understand this issue is to change from the risk-neutral measure $\mathbb{Q}$ to the equivalent T-forward measure defined as follows:

Definition 7.1. *The T-forward measure $\mathbb{Q}^T$ is the measure equivalent to $\mathbb{Q}$ with Radon–Nykodym density*

$$\frac{d\mathbb{Q}^T}{d\mathbb{Q}} = \frac{\exp\left(-\int_0^T r_s ds\right)}{P(0,T)}.$$

By Bayes' formula, the price of the claim ξ is given by

$$\begin{aligned} V_t = \mathbb{E}^{\mathbb{Q}}\{e^{-\int_t^T r_s ds}\xi|\mathcal{F}_t\} &= \mathbb{E}^{\mathbb{Q}^T}\{e^{\int_0^t r_s ds}P(0,T)\xi|\mathcal{F}_t\}\mathbb{E}^{\mathbb{Q}}\{e^{-\int_t^T r_s ds}P(0,T)^{-1}|\mathcal{F}_t\} \\ &= P(t,T)\mathbb{E}^{\mathbb{Q}^T}\{\xi|\mathcal{F}_t\}. \end{aligned}$$

In particular, we have the product representation of the price which we could only derive under $\mathbb{P}$-conditional independence. The forward measure $\mathbb{Q}^T$ combines the discount factor with the risk-neutral pricing measure; consequently, the measures $\mathbb{Q}^S$ and $\mathbb{Q}^T$ are generally different if $T \neq S$, unless r_t is deterministic.

The forward rate $\{f(t,T)\}_{t\in[0,T]}$ is a martingale under the T-forward measure $\mathbb{Q}^T$ since we have the formal calculation

$$\begin{aligned} f(t,T) &= -\frac{1}{P(t,T)}\frac{\partial}{\partial T}P(t,T) \\ &= -\frac{1}{P(t,T)}\frac{\partial}{\partial T}\mathbb{E}^{\mathbb{Q}}\left\{e^{-\int_t^T r_s ds}|\mathcal{F}_t\right\} \\ &= \frac{1}{P(t,T)}\mathbb{E}^{\mathbb{Q}}\left\{r_T e^{-\int_t^T r_s ds}|\mathcal{F}_t\right\} \\ &= \mathbb{E}^{\mathbb{Q}^T}\{r_T|\mathcal{F}_t\}. \end{aligned}$$

The above interchange of differentiation and expectation is justified if the short rate is positive or sufficiently well-behaved.

Now let us put ourselves in the framework of an abstract HJM model as studied in Chap. 6. In particular, assume that the probability space $(\Omega, \mathcal{F}, \mathbb{Q})$ supports a Wiener process $\{W_t\}_{t\geq 0}$ defined cylindrically on a Hilbert space G, and that the risk-neutral dynamics of the forward rate process $\{f_t\}_{t\geq 0}$ are given by

$$df_t = (Af_t + F_{\mathrm{HJM}}(\sigma_t))dt + \sigma_t dW_t$$

where A is the generator of the shift semigroup $\{S_t\}_{t\geq 0}$ on the state space F, and $\{\sigma_t\}_{t\geq 0}$ is an adapted process valued in the space $\mathcal{L}_{\mathrm{HS}}(G, F)$ of Hilbert–Schmidt operators taking G into F. The forward rates solve the integral equation

$$f_t = S_t f_0 + \int_0^t S_{t-s} F_{\mathrm{HJM}}(\sigma_s) ds + \int_0^t S_{t-s}\sigma_s dW_s$$

and in particular, the forward rate $f(t,T) = f_t(T-t)$ is given by

$$\begin{aligned} f(t,T) &= f(0,T) + \int_0^t \langle \sigma_s^* \delta_{T-s}, \sigma_s^* I_{T-s}\rangle ds + \int_0^t S_{t-s}\sigma_s^* \delta_{T-s} dW_s \\ &= f(0,T) + \int_0^t \sigma_s^* I_{T-s}\left(dW_s + \sigma_s^* \delta_{T-s} ds\right) \end{aligned}$$

where δ_x is the evaluation functional $\delta_x(f) = f(x)$ and I_x is the definite integral functional $I_x(f) = \int_0^x f(s)ds$. We now see how to change from the risk-neutral measure $\mathbb{Q}$ to the forward measure $\mathbb{Q}^T$: since the forward rate $\{f(t,T)\}_{t\in[0,T]}$ is a $\mathbb{Q}^T$-martingale, we must have that

$$W_t^T = W_t + \int_0^t \sigma_s^* \delta_{T-s} ds$$

defines a cylindrical $\mathbb{Q}^T$ Wiener process. Formalizing the proceeding discussion leads to the following theorem:

Theorem 7.3. *Let (σ, λ) be an abstract HJM model on a Hilbert space F, and let $\mathbb{Q}$ be the measure on $(\Omega, \mathcal{F})$ such that the discounted prices of all the bonds are martingales. Then the measure $\mathbb{Q}^T$ with density*

$$\frac{d\mathbb{Q}^T}{d\mathbb{Q}} = \exp\left(-\frac{1}{2}\int_0^T \|\sigma_s^* I_{T-s}\|_G^2 + \int_0^T \sigma_s^* I_{T-s} dW_s\right)$$

is the T-forward measure. The process $\{W_t^T\}_{t\in[0,T]}$ defined cylindrically on G by the formula

$$W_t^T = W_t + \int_0^t \sigma_s^* I_{T-s} ds$$

is a Wiener process for the forward measure $\mathbb{Q}^T$.

7.3.2 LIBOR Rates Revisited

Shifting away from the HJM framework with its emphasis on continuously compounded forward rates, we now come back to the discretely compounded interest rates already discussed in the first chapter. The LIBOR rate is defined by the formula

$$L(t,T) = \frac{1}{\delta}\left(\frac{P(t,T)}{P(t,T+\delta)} - 1\right).$$

It is the value at time $t \geq 0$ for the simple interest accumulated from T to $T+\delta$ where $\delta > 0$ is a fixed period of time, typically 3 months. Note that as $\delta \searrow 0$ the LIBOR rate approaches the forward rate $f(t,T)$.

Notice that the LIBOR rate process $\{L(t,T)\}_{t\in[0,T]}$ is a martingale for the $T+\delta$- forward measure. Indeed, just observe that

$$\begin{aligned}
1 + \delta L(t,T) &= \frac{P(t,T)}{P(t,T+\delta)} \\
&= \frac{\mathbb{E}^{\mathbb{Q}}\{e^{-\int_t^T r_s ds}|\mathcal{F}_t\}}{P(t,T+\delta)} \\
&= \mathbb{E}^{\mathbb{Q}^{T+\delta}}\{e^{\int_T^{T+\delta} r_s ds}|\mathcal{F}_t\}.
\end{aligned}$$

In the modeling framework proposed by Brace, Gratarek, and Musiela in [24] the LIBOR rate is taken as the primitive state process. In particular, for each maturity date $T \geq 0$ the dynamics of the LIBOR rate $\{L(t,T)\}_{t\in[0,T]}$ are assumed to be log-normal under the corresponding $T+\delta$ forward measure.

That is, assume that there is a function $\gamma : \mathbb{R}_+^2 \to G$ such that for each $T \geq 0$, we have

$$dL(t,T) = L(t,T)\gamma(t,T)dW_t^{T+\delta}.$$

The advantage of such a modeling assumption is that the prices of LIBOR contingent claims with payouts of the form $\xi = g(L(T,T))$ which settle in arrears (i.e. the money changes hands at the date $T+\delta$) are very easy to compute:

$$\begin{aligned}
V_t &= \mathbb{E}^{\mathbb{Q}}\{e^{-\int_t^{T+\delta} r_s ds} g(L(T,T))|\mathcal{F}_t\} \\
&= P(t,T+\delta)\mathbb{E}^{\mathbb{Q}^{T+\delta}}\{g(L(T,T)|\mathcal{F}_t\} \\
&= P(t,T+\delta)\int_{-\infty}^{\infty} g(L(t,T)e^{\sigma_0\sqrt{T-t}z - \sigma_0^2(T-t)/2})\frac{e^{-z^2/2}}{\sqrt{2\pi}}dz
\end{aligned}$$

where

$$\sigma_0 = \left(\frac{1}{T-t}\int_t^T \|\gamma(s,T)\|^2 ds\right)^{1/2}$$

is the effective volatility. In particular, if the claim is a caplet then the corresponding payout function is $g(x) = (x-K)^+$ where K is the strike price, and hence the price of a caplet in the BGM framework is given by the familiar Black–Scholes formula:

$$L(t,T)\Phi(d_1) - \kappa\Phi(d_2)$$

where Φ is the standard normal distribution function and

$$d_1 = \frac{\log(L(t,T)/K) + \sigma_0^2(T-t)/2}{\sigma_0\sqrt{T-t}} \text{ and } d_2 = \frac{\log(L(t,T)/K) - \sigma_0^2(T-t)/2}{\sigma_0\sqrt{T-t}}.$$

Of course, the benefit of such a pricing formula is that practitioners are very comfortable with the Black–Scholes formula, and indeed pricing caplets in this manner seems to be the market practice.

The question arises: Can a BGM model for the LIBOR rates be specified for all of the maturities T simultaneously? Furthermore, can it be recast as an abstract HJM model studied in Chap. 6? Fortunately, Brace, Gatarek, and Musiela proved that the log-normal modeling assumptions are mutually compatible for different maturity dates T.

We now explore the dynamics of the forward rates under the risk-neutral measure $\mathbb{Q}$. The bond prices are given by

$$P(t,T) = P(0,T)\exp\left(\int_0^t r_s ds - \frac{1}{2}\int_0^t \|\sigma_s^* I_{T-s}\|^2 duds - \int_0^t \sigma_s^* I_{T-s} dW_s\right)$$

and hence by the BGM assumption we have

$$\sigma_t^*(I_{x+\delta} - I_x) = \frac{\delta L(t,t+x)}{1+\delta L(t,t+x)}\gamma(t,t+x).$$

If we set $\sigma_t^*\delta_x = 0$ for $0 \le x \le \delta$, then summing the above equation shows that the integrated volatility is given by

$$\sigma_t^* I_x = \sum_{k=1}^{[x/\delta]} \frac{\delta L(t,t+x-\delta k)}{1+\delta L(t,t+x-k\delta)}\gamma(t,t+x-k\delta).$$

If γ is sufficiently smooth in its second argument, the HJM volatility can be recovered by differentiating both sides of the above equation with respect to x.

Notes & Complements

Vargiolu [131] studied the risk-neutral dynamics of the linear Gaussian HJM model when the state space is one of the family of Sobolev spaces $\{H_\gamma^1\}_{\gamma\ge 0}$ defined by

$$H_\gamma^1 = \left\{f : \mathbb{R}_+ \to \mathbb{R}_+; \int_0^\infty (f(x)^2 + f'(x)^2)e^{-\gamma x}dx\right\}.$$

In this case there exists a mild solution to equation the HJM equation in the space H^1_γ if

$$\sum_{i=1}^{\infty} \|\sigma^i\|^2_{H^1_\gamma} + \|\sigma^i\|^4_{H^1_0} + \|\sigma^i\|^4_{L^4_\gamma} < +\infty.$$

If in addition $\sum_{i=1}^{\infty} \|\sigma^i\|^2_{H^1_0} < +\infty$ then there exists a family of Gaussian invariant measures on H^1_γ. Tehranchi [129] studied the existence of invariant measures for general Markovian HJM models in the space H_w. A major difference between Vargiolu's space H^1_γ and the space H_w used in the analysis of this chapter is that on H^1_γ the semigroup $\{S_t\}_{t\geq 0}$ has many non-trivial invariant measures.

We learned from Rogers the simple proof presented in this chapter that the average short rate is a maximum, where the average is taken in an invariant measure of the risk-neutral HJM dynamics.

Motivated by specific problems in fixed income security portfolio risk management, Bouchaud, Cont, El Karoui, Potters, and Sagna propose a model for the dynamics of the forward curve under the historical probability measure $\mathbb{P}$. See [23], [22] and [41]. They first model the short and the long rates as a bivariate diffusion, and then, they consider a random string model for the deviation of the forward curve from the straight line joining the short to the long rate. Since they do not have to worry about the drift condition, they propose to add a regularizing term in the drift, in the form of a perturbation of the shift operator A by a Laplacian term.

The parabolic SPDE model of the deformation process, as discussed in Sect. 7.2 is closely related to the stochastic cable equation studied by Walsh [132, 133] as a model neural response. The solutions of this parabolic SPDE driven by space–time white noise is very different in dimensions greater than one. In particular, the solutions are not functions, and have to be interpreted as random distributions. Independently of the financial motivation which seems to have something to do with arbitrage opportunities created by transaction costs, the main effect of this perturbation is to replace a hyperbolic SPDE by a parabolic one. Once the operator A includes a Laplacian component, it is easy to choose the Hilbert space F to make sure that this operator is self-adjoint, and because of the properties of the Laplacian, this operator A becomes strictly negative, and the solution of the SPDE is an infinite dimensional Ornstein–Uhlenbeck process of the best kind since it has a unique invariant measure whose covariance structure can be identified with the computations given in this book. In this way, the PCA can be performed both in the model and on the data and the quality of the fit can be assessed in an empirical manner. Our understanding is that work is in process to quantify this fit.

The BGM model of the LIBOR rates was introduced by Brace, Gatarek, and Musiela [24]. These authors proved that such a model exists, and that the model dynamics exhibit an invariant measure on the space of continuous functions.

References

1. R.J. Adler. *An introduction to continuity, extrema, and related topics for general Gaussian processes*. Institute of Mathematical Statistics Lecture Notes – Monograph Series, 12. Institute of Mathematical Statistics, Hayward, CA, 1990.
2. Y. Ait-Sahalia. Do interest rates really follow continuous time Markov diffusions?, 1997.
3. A. Antoniadis and R. Carmona. Eigenfunction expansions for infinite-dimensional Ornstein-Uhlenbeck processes. *Probab. Theory Related Fields*, 74(1):31–54, 1987.
4. M. Avellaneda and A. Majda. Mathematical models with exact renormalization for turbulent transport. *Comm. Math. Phys.*, 131(2):381–429, 1990.
5. F. Baudoin and J. Teichmann. Hypo-ellipticity in infinite dimensions. *Annals Appl. Probab.*, 15 (3):1765–1777, 2005.
6. C.A. Bester. Random field and affine models for interest rates: An empirical comparison, 2004.
7. T.R. Bielecki and M. Rutkowski. *Credit Risk: Modelling, Valuation and Hedging*. Springer Finance. Springer-Verlag, Berlin Heidelberg, 2002.
8. T. Björk. Interest rate theory. In W. Runggaldier, editor, *Financial Mathematics (Bressanone, 1996)*, volume 1656 of *Lecture Notes in Math.*, pages 53–122. Springer-Verlag, Berlin Heidelberg, 1997.
9. T. Björk. A geometric view of interest rate theory. In *Option Pricing, Interest Rates and Risk Management*, Handb. Math. Finance, pages 241–277. Cambridge Univ. Press, Cambridge, 2001.
10. T. Björk. On the geometry of interest rate models. In R. Carmona et. al., editor, *Paris-Princeton Lectures on Mathematical Finance 2003*, volume 1847 of *Lecture Notes in Math.*, pages 133–215. Springer-Verlag, Berlin Heidelberg, 2004.
11. T. Björk and B.J. Christensen. Interest rate dynamics and consistent forward rate curves. *Math. Finance*, 9(4):323–348, 1999.
12. T. Björk and A. Gombani. Minimal realizations of interest rate models. *Finance Stoch.*, 3(4):413–432, 1999.
13. T. Björk, Y. Kabanov, and W. Runggaldier. Bond market structure in the presence of marked point processes. *Math. Finance*, 7(2):211–239, 1997.

14. T. Björk and C. Landén. On the construction of finite dimensional realizations for nonlinear forward rate models. *Finance Stoch.*, 6(3):303–331, 2002.
15. T. Björk, C. Landén, and L. Svensson. Finite-dimensional Markovian realizations for stochastic volatility forward-rate models. *Proc. R. Soc. Lond. Ser. A Math. Phys. Eng. Sci.*, 460(2041):53–83, 2004. Stochastic analysis with applications to mathematical finance.
16. T. Björk, G. Di Masi, Y. Kabanov, and W. Runggaldier. Towards a general theory of bond markets. *Finance and Stochastics*, 1:141–174, 1997.
17. T. Björk and L. Svensson. On the existence of finite-dimensional realizations for nonlinear forward rate models. *Math. Finance*, 11(2):205–243, 2001.
18. V. Bogachev. *Gaussian Measures.* Mathematical Surveys and Monographs.
19. R. Bonic and J. Frampton. Differentiable fucntions on certain banach spaces. *Bull. Amer. Math. Soc.*, 71:393–395, 1965.
20. B. Bouchard, I. Ekeland, and N. Touzi. On the Malliavin approach to Monte Carlo approximation of conditional expectations. *Finance Stoch.*, 8(1):45–71, 2004.
21. B. Bouchard and N. Touzi. Discrete-time approximation and Monte-Carlo simulation of backward stochastic differential equations. *Stochastic Process. Appl.*, 111(2):175–206, 2004.
22. J.P. Bouchaud, R. Cont, N. El Karoui, M. Potters, and N. Sagna. Strings attached. *RISK*, Jul 1998.
23. J.P. Bouchaud, R. Cont, N. El Karoui, M. Potters, and N. Sagna. Phenomenology of the interest rate curve. Technical report, 1999.
24. A. Brace, D. Gatarek, and M. Musiela. The market model of interest rate dynamics. *Math. Finance*, 7(2):127–155, 1997.
25. D. Brigo and F. Mercurio. *Interest Rate Models.* Springer-Verlag, 2001.
26. R.H. Cameron and W.T. Martin. Transformations of Wiener integrals under translations. *Annals of Mathematics*, 45(2):389–396, 1944.
27. M. Capitaine, E.P. Hsu, and M. Ledoux. Martingale representation and a simple proof of logarithmic Sobolev inequalities on path spaces. *Electron. Comm. Probab.*, 2:71–81 (electronic), 1997.
28. R. Carmona. Measurable norms and some Banach space valued Gaussian processes. *Duke Math. J.*, 44(1):109–127, 1977.
29. R. Carmona. Tensor products of Gaussian measures. In *Proc. Conf. Vector Space Measures and Applications, Dublin*, number 644 in Lect. Notes in Math., pages 96–124, 1977.
30. R. Carmona. Infinite dimensional Newtonian potentials. In *Proc. Conf. on Probability Theory on Vector Spaces II, Wroclaw (Poland)*, volume 828 of *Lect. Notes in Math.*, pages 30–43, 1979.
31. R. Carmona. Transport properties of Gaussian velocity fields. In M.M. Rao, editor, *Real and Stochastic Analysis: Recent Advances*, Probab. Stochastics Ser., pages 9–63. CRC, Boca Raton, FL, 1997.
32. R. Carmona. *Statistical Analysis of Financial Data in* `S-Plus`. Springer Texts in Statistics. Springer-Verlag, New York, 2004.
33. R. Carmona and F. Cerou. Transport by incompressible random velocity fields: simulations & mathematical conjectures. In R.A. Carmona & B. Rozovskii, editor, *Stochastic partial differential equations: six perspectives*, volume 64 of *Math. Surveys Monogr.*, pages 153–181. Amer. Math. Soc., Providence, RI, 1999.

34. R. Carmona and S. Chevet. Tensor Gaussian measures on $L^p(E)$. *J. Funct. Anal.*, 33(3):297–310, 1979.
35. R. Carmona and J. Lacroix. *Spectral Theory of Random Schrödinger Operators.* Probability and its Applications. Birkhäuser Boston Inc., Boston, MA, 1990.
36. R. Carmona and M. Tehranchi. A characterization of hedging portfolios for interest rate contingent claims. *Ann. Appl. Probab.*, 14(3):1267–1294, 2004.
37. R. Carmona and L. Wang. Monte Carlo Malliavin computation of the sensitivities of solutions of SPDE's, 2005.
38. S. Chevet. Un résultat sur les mesures gaussiennes. *C. R. Acad. Sci. Paris Sér. A-B*, 284(8):A441–A444, 1977.
39. J.M.C. Clark. The representation of functionals of Brownian motion by stochastic integrals. *Ann. Math. Statist.*, 41:1282–1295, 1970.
40. P. Collin-Dufresne and R.S. Goldstein. Generalizing the affine framework to hjm and random field models. Technical report, Carnegie Mellon, 2003.
41. Rama Cont. Modeling term structure dynamics: an infinite dimensional approach. *Int. J. Theor. Appl. Finance*, 8(3):357–380, 2005.
42. J. Cvitanic, J. Ma, and J. Zhang. Efficient computation of hedging portfolios for options with discontinuous payoffs. *Mathematical Finance*, 13:135–151, 2003.
43. G. Da Prato and J. Zabczyk. *Stochastic Equations in Infinite Dimensions*, volume 44 of *Encyclopedia of Mathematics and its Applications.* Cambridge University Press, Cambridge, 1992.
44. Ju.L. Daleckiĭ and S.V. Fomin. Differential equations for distributions in infinite-dimensional spaces. *Trudy Sem. Petrovsk.*, (4):45–64, 1978.
45. D.A. Dawson. Stochastic evolution equations. *Math. Bio. Sci.*, 15:287–316, 1972.
46. D.A. Dawson. Stochastic evolution equations and related measure processes. *J. Multivariate Anal.*, 5:1–52, 1975.
47. M. De Donno. A note on completeness in large financial markets. *Math. Finance*, 14(2):295–315, 2004.
48. M. De Donno and M. Pratelli. On the use of measured-valued strategies in bond markets. *Finance Stoch.*, 8:87–109, 2004.
49. D. Duffie, D. Filipović, and W. Schachermayer. Affine processes and applications in finance. *Ann. Appl. Probab.*, 13(3):984–1053, 2003.
50. D. Duffie and R. Kan. Multifactor models of the term structure. In *Mathematical Models in Finance.* Chapman & Hall, London, 1995.
51. D. Duffie and K. Singleton. *Credit Risk.* Princeton University Press, Princeton, NJ, 2003.
52. P. Dybvig, J. Ingersoll, and S. Ross. Long forward and zero-coupon rates can never fall. *J. Business*, 69:1–12, 1996.
53. I. Ekeland and E. Taflin. A theory of bond portfolios. *Annals of Applied Probability*, 15:1260–1305, 2002.
54. R. Elie, J.D. Fermanian, and N. Touzi. Optimal greek weights by kernel estimates, April 2005.
55. R.J. Elliott and J. van der Hoek. Stochastic flows and the forward measure. *Finance Stochast.*, 5:511–525, 2001.
56. F. Fabozzi. *The Handbook of Fixed Income Securities.* McGraw Hill, 6th edition, 2000.

57. F. Fabozzi. *The Handbook of Mortgage Backed Securities.* McGraw Hill, 2001.
58. J. Feldman. Equivalence and perpendicularity of Gaussian measures. *Pacific Journal of Mathematics*, 9:699–708, 1958.
59. W. Feller. *An Introduction to Probability Theory and its Applications*, volume II. Wiley & Sons, New York, NY, 1971.
60. X. Fernique. Régularité des trajectoires des fonctions aléatoires gaussiennes. In *École d'Été de Probabilités de Saint-Flour, IV-1974*, volume 480 of *Lecture Notes in Math.*, pages 1–96. Springer-Verlag, Berlin Heidelberg, 1975.
61. D. Filipović. Invariant manifolds for weak solutions to stochastic equations. *Probab. Theory Related Fields*, 118(3):323–341, 2000.
62. D. Filipović. *Consistency Problems for Heath-Jarrow-Morton Interest Rate Models*, volume 1760 of *Lecture Notes in Mathematics.* Springer-Verlag, Berlin Heidelberg, 2001.
63. D. Filipović and J. Teichmann. Existence of invariant manifolds for stochastic equations in infinite dimension. *J. Funct. Anal.*, 197(2):398–432, 2003.
64. D. Filipović and J. Teichmann. Regularity of finite-dimensional realizations for evolution equations. *J. Funct. Anal.*, 197(2):433–446, 2003.
65. D. Filipović and J. Teichmann. On the geometry of the term structure of interest rates. *Proc. R. Soc. Lond. Ser. A Math. Phys. Eng. Sci.*, 460(2041):129–167, 2004.
66. E. Fournié, J.-M. Lasry, J. Lebuchoux, and P.-L. Lions. Applications of Malliavin calculus to Monte-Carlo methods in finance. II. *Finance Stoch.*, 5(2):201–236, 2001.
67. E. Fournié, J.-M. Lasry, J. Lebuchoux, P.-L. Lions, and N. Touzi. Applications of Malliavin calculus to Monte Carlo methods in finance. *Finance Stoch.*, 3(4):391–412, 1999.
68. B. Gaveau. Intégrale stochastique radonifiante. *C. R. Acad. Sci. Paris*, ser. A 276, May 1973.
69. E. Gobet and A. Kohatsu-Higa. Computation of Greeks for barrier and lookback options using Malliavin calculus. *Electron. Comm. Probab.*, 8:51–62 (electronic), 2003.
70. R. Goldstein. The term structure of interest rates as a random field. *Review of Financial Studies*, 13(2):365–384, 2000.
71. V. Goodman. Distribution estimates for functionals of the two-parameter wiener process. *Ann. Probab.*, 4(6):977–982, 1976.
72. V. Goodman. Quasi-differentiable functions of Banach spaces. *Proc. Amer. Math. Soc.*, 30:367–370, 1971.
73. L. Gross. Measurable functions on Hilbert space. *Trans. Amer. Math. Soc.*, 105:372–390, 1962.
74. L. Gross. Abstract Wiener spaces. In *Proc. Fifth Berkeley Sympos. Math. Statist. and Probability (Berkeley, Calif., 1965/66), Vol. II: Contributions to Probability Theory, Part 1*, pages 31–42. Univ. California Press, Berkeley, Calif., 1967.
75. L. Gross. Potential theory on Hilbert space. *J. Functional Analysis*, 1:123–181, 1967.
76. L. Gross. Logarithmic Sobolev inequalities. *Amer. J. Math.*, 97(4):1061–1083, 1975.
77. L. Gross. On the formula of Mathews and Salam. *J. Functional Analysis*, 25(2):162–209, 1977.

78. J. Hájek. On a property of normal distributions of any stochastic process. *Czechoslovak Mathematical Journal*, 8:610–617, 1958.
79. A.T. Hansen and R. Poulsen. A simple regime switching term structure model. *Finance and Stochastics*, 4:409–429, 2000.
80. D. Heath, R. Jarrow, and A. Morton. Bond pricing and the term structure of interest rates: A new methodology for contingent claims valuation. *Econometrica*, 60:77–105, 1992.
81. R. Holley and D.W. Stroock. Generalized Ornstein Ulhenbeck processes and infinite particle branching brownian motions. *Publ. RIMS Kyoto Univ.*, 14:741–788, 1978.
82. F. Hubalek, I. Klein, and J. Teichmann. A general proof of the Dybvig-Ingersoll-Ross theorem: long forward rates can never fall. *Math. Finance*, 12(4):447–451, 2002.
83. F. Jamshidian. Bond, futures and option evaluation in the quadratic interest rate model. *Applied Mathematical Finance*, 3:93–115, 1996.
84. J. Kiefer J.R. Blum and M. Rosenblatt. Distribution free tests of independence based on the sample distribution function. *Ann. Math. Statist.*, 32:485–498, 1961.
85. I. Karatzas, D.L. Ocone, and J. Li. An extension of Clark's formula. *Stochastics Stochastics Rep.*, 37(3):127–131, 1991.
86. I. Karatzas and S.E. Shreve. *Brownian Motion and Stochastic Calculus*, volume 113 of *Graduate Texts in Mathematics*. Springer-Verlag, New York, second edition, 1991.
87. N. El Karoui, R. Mynemi, and R. Viswanathan. Arbitrage pricing and hedging of interest rate claims with state variables: theory and applications. Technical report, Stanford Univ. and Paris VI, 1992.
88. D.P. Kennedy. The term structure of interest rates as a Gaussian random field. *Math. Finance*, 4(3):247–258, 1994.
89. D.P. Kennedy. Characterizing Gaussian models of the term structure of interest rates. *Math. Finance*, 7:107–118, 1997.
90. R.L. Kimmel. Modeling the term structure of interest rates: A new approach. *Journal of Financial Economics*, 72:143–183, 2004.
91. A. Kohatsu-Higa and M. Montero. Malliavin calculus in finance. In *Handbook of Computational and Numerical Methods in Finance*, pages 111–174. Birkhäuser Boston, Boston, MA, 2004.
92. A. Kriegl and P.W. Michor. *The Convenient Setting of Global Analysis*, volume 53 of *Mathematical Surveys and Monographs*. American Mathematical Society, Providence, RI, 1997.
93. J. Kuelbs. The law of the iterated logarithm and related strong convergence theorems for banach space valued random variables. In *Ecole d'Eté de Probabilités de Saint Flour 1975*, volume 539 of *Lect. Notes in Math.*, pages 224–314. Springer-Verlag, New York, NY, 1976.
94. H.H. Kuo. *Gaussian Measure on Banach Space*. Lect. Notes in Math. 1975.
95. J. Kurzweil. On approximation in real banach spaces. *Studia Math.*, 14:213–231, 1954.
96. S. Kusuoka. Term structure and SPDE. In *Advances in Mathematical Economics*, pages 67–85. Springer-Verlag, New York, NY, 2000.
97. D. Lamberton and B. Lapeyre. *Introduction to Stochastic Calculus Applied to Finance*. Chapman & Hall, London, 1996. Translated from the 1991 French original by Nicolas Rabeau and François Mantion.

98. H.J. Landau and L.A. Shepp. On the supremum of a Gaussian process. *Sankhyā Ser. A*, 32:369–378, 1970.
99. D. Lando. *Credit Risk Modeling: Theory and Applications.* Princeton University Press, Princeton, NJ, 2004.
100. M. Ledoux and M. Talagrand. *Probability in Banach Spaces*, volume 23 of *Ergebnisse der Mathematik und ihrer Grenzgebiete (3) [Results in Mathematics and Related Areas (3)].* Springer-Verlag, Berlin Heidelberg, 1991. Isoperimetry and processes.
101. R. Litterman and J. Scheinkman. Common factors affecting bond returns. *J. of Fixed Income*, 1:49–53, 1991.
102. F.A. Longstaff and R.S. Schwartz. Valuing american options by simulation: A simple least-square approach. *Review of Financial Studies*, 14:113–147, 2001.
103. N. Touzi M. Mrad and A. Zeghal. Monte-carlo estimation of a joint density using malliavin calculus and application to american options.
104. P.A. Meyer. Infinite dimensional Ornstein Uhlenbeck process. In *Séminaire de Probabilités***XXX**, Lect. Notes in Math. Springer-Verlag, New York, NY, 1990.
105. R.A. Minlos. Generalized random processes and their extension to a measure. In *Selected Transl. Math. Statist. and Prob., Vol. 3*, pages 291–313. Amer. Math. Soc., Providence, R.I., 1963.
106. M. Musiela and B. Goldys. Infinite dimensional diffusions, Kolmogorov equations and interest rate models. In E. Jouini, J. Cvitanic, and M. Musiela, editors, *Option Pricing, Interest Rates and Risk Management*, Handb. Math. Finance, pages 314–345. Cambridge Univ. Press, Cambridge, 2001.
107. M. Musiela and M. Rutkowski. *Martingale Methods in Financial Modelling.* Springer-Verlag, 1997.
108. E. Nelson. The free Markov field. *J. Functional Anal.*, 12:211–227, 1973.
109. D. Nualart. *The Malliavin Calculus and Related Topics.* Springer-Verlag, New York, NY, 1995.
110. D. Ocone. Malliavin's calculus and stochastic integral representations of functionals of diffusion processes. *Stochastics*, 12(3-4):161–185, 1984.
111. D.L. Ocone and I. Karatzas. A generalized Clark representation formula, with application to optimal portfolios. *Stochastics Rep.*, 34(3-4):187–220, 1991.
112. M.A. Piech. The Ornstein-Uhlenbeck semigroup in an infinite dimensional L^2 setting. *J. Functional Analysis*, 18:271–285, 1975.
113. G. Da Prato and J. Zabczyk. *Ergodicity for Infinite Dimensional Systems*, volume 229 of *Lecture Notes Series.* Cambridge University Press, 1996.
114. R. Rebonato. *Modern Pricing of Interest-Rate Derivatives: the LIZBOR Market Model and Beyond.* Princeton University Press, 1992.
115. R. Rebonato. *Interest-Rate Option Models: Understanding, Analyzing and Using Models for Exotic Interest-Rate Options.* Wiley & Sons, 2nd edition, 1998.
116. N. Ringer and M. Tehranchi. Optimal portfolio choice in bond markets, submitted for publication 2005.
117. L.C.G. Rogers. The potential approach to the term structure of interest rates and foreign exchange rates. *Math. Finance*, 7(2):157–176, 1997.
118. L.C.G. Rogers and D. Williams. *Diffusions, Markov processes, and martingales. Volume 2: Itô Calculus.* Cambridge Mathematical Library. Cambridge University Press, Cambridge, 2nd edition, 2000.

119. A. Roncoroni and P. Guiotto. Theory and calibration of HJM with shape factors. In *Mathematical finance – Bachelier Congress, 2000 (Paris)*, Springer Finance, pages 407–426. Springer-Verlag, Berlin Heidelberg, 2002.
120. A. Roncoroni, P. Guiotto, and S. Galluccio. Shape factors and cross-sectional risk. *(preprint)*, 2005.
121. H. Satô. Gaussian measure on a Banach space and abstract Wiener measure. *Nagoya Math. J.*, 36:65–81, 1969.
122. P. Schönbucher. *Credit Derivatives Pricing Models*. Wiley & Sons, 2003.
123. L. Schwartz. Mesures cylindriques et applications radonifiantes dans les espaces de suites. In *Proc. Internat. Conf. on Functional Analysis and Related Topics (Tokyo, 1969)*, pages 41–59. Univ. of Tokyo Press, Tokyo, 1970.
124. H. Shirakawa. Interest rate options pricing with Poisson-Gaussian forward rate curve processes. *Mathematical Finance*, 1:77–94, 1991.
125. A.V. Skorohod. *Random Linear Operators*. Mathematics and its Applications (Soviet Series). D. Reidel Publishing Co., Dordrecht, 1984. Translated from the Russian.
126. O.G. Smoljanov and S.V. Fomīn. Measures on topological linear spaces. *Usephi Mat. Nauk*, 31(4, (190)):3–56, 1976.
127. D.W. Stroock. The Malliavin calculus and its application to second order parabolic differential equations. I. *Math. Systems Theory*, 14(1):25–65, 1981.
128. E. Taflin. Bond market completeness and attainable contingent claims. *Finance and Stochastics*, 4(3), 2005.
129. M. Tehranchi. A note on invariant measures for HJM models. *Finance and Stochastics*, 4(3), 2005.
130. J.N. Tsitsiklis and B. Van Roy. Optimal stopping of markov processes: Hilbert space theory, approximation algorithms, and an application to pricing high-dimensional financial derivatives. *IEEE Trans. Automat. Control*, 44 (10):1840–1851, 1999.
131. T. Vargiolu. Invariant measures for the Musiela equation with deterministic diffusion term. *Finance Stoch.*, 3(4):483–492, 1999.
132. J.B. Walsh. A stochastic model of neural response. *Adv. in Appl. Probab.*, 13(2):231–281, 1981.
133. J.B. Walsh. An introduction to stochastic partial differential equations. In *École d'été de probabilités de Saint-Flour, XIV – 1984*, volume 1180 of *Lecture Notes in Math.*, pages 265–439. Springer-Verlag, Berlin Heidelberg, 1986.
134. J.H.M. Whittfield. Differentiable functions with bounded non-empty support on banach spaces. *Bull. Amer. Math. Soc.*, 71:145–146, 1966.
135. J. Yeh. Wiener measure in a space of functions of two variables. *Trans. Amer. Math. Soc.*, 95:433–450, 1960.
136. M. Yor. Existence et unicité de diffusions à valeurs dans un espace de Hilbert. *Ann. Inst. H. Poincaré Sect. B (N.S.)*, 10:55–88, 1974.
137. J. Zabczyk. Stochastic invariance and consistency of financial models. *Rend. Mat. Acc. Lincei*, 9:67–80, 2000.

Notation Index

Author Index

Subject Index

MIX
Papier aus verantwortungsvollen Quellen
Paper from responsible sources
FSC® C105338

If you have any concerns about our products,
you can contact us on
ProductSafety@springernature.com

In case Publisher is established outside the EU,
the EU authorized representative is:
Springer Nature Customer Service Center GmbH
Europaplatz 3, 69115 Heidelberg, Germany

Printed by Libri Plureos GmbH
in Hamburg, Germany